Maximizing Broadband Services to Rural Communities

MAXIMIZING BROADBAND SERVICES TO RURAL COMMUNITIES

VICTOR B. HAYNESWORTH
AND
ALAN R. HARRISON
EDITORS

Nova Science Publishers, Inc.
New York

For permission to use material from this book please contact us:
Telephone 631-231-7269; Fax 631-231-8175
Web Site: http://www.novapublishers.com

NOTICE TO THE READER

LIBRARY OF CONGRESS CATALOGING-IN-PUBLICATION DATA
Available upon request

ISBN 978-1-60692-846-2

Published by Nova Science Publishers, Inc. ✢ New York

CONTENTS

PREFACE

This book focuses on the challenges for providing reliable, affordable broadband access of the rural small businesses. Experience has shown that broadband can bring economic revitalization to small towns by creating clusters of small businesses. Broadband service can also help farmers and farm-related businesses control costs and optimize production. This technology can provide real-time access to whether reports, fertilization guidance, and livestock tracking. Farming communities must maximum the use of high-tech Internet access to ensure further development. The authors are concerned that many of our country's rural and agricultural-based communities are not yet fully realizing the potential of value of broadband services to their economies. As more of these communities gain highspeed access to the Internet, the next challenge is to help them use the access effectively to help create jobs and sustain growth. There are many debates going on right now about broadband policy.

Chapter 1 - A USDA Rural Development is committed to fulfilling the President and Congress' goal of bringing affordable broadband to all rural Americans. This is a high priority. Broadband is a transformative technology. It allows rural communities to dramatically enhance the quality of health care and education. It gives every rural business access to regional, national, and international markets. It reduces barriers of time and distance, levels the playing field, and makes rural communities better places to live, work, and raise a family.

Rural America is diverse and so is the broadband program. USDA Rural Development tailors solutions to the needs of individual communities. We have financed technologies including fiber to the home, DSL, wireless, broadband over power lines, and hybrid fiber/ coax systems. We are eager to work with startup companies as well as established providers and stand ready to consider innovative financing arrangements. Our mission is to increase economic opportunity and improve the quality of life in rural America. Broadband is essential to both of these goals, and we will not rest until the job is done.

Chapter 2 - Broadband or high-speed Internet access is provided by a series of technologies that give users the ability to send and receive data at volumes and speeds far greater than current Internet access over traditional telephone lines. In addition to offering speed, broadband access provides a continuous, "always on" connection and the ability to both receive (download) and transmit (upload) data at high speeds. Broadband access, along with the content and services it might enable, has the potential to transform the Internet: both what it offers and how it is used. It is possible that many of the future applications that will best exploit the technological capabilities of broadband have yet to be developed. There are multiple transmission media or technologies that can be used to provide broadband access. These include cable; an enhanced telephone service called digital subscriber line (DSL); fiber-tothe-home (FTTH); satellite, mobile, and fixed wireless (including "wi-fi" and "WiMax"); broadband over powerlines (BPL); and others.

From a public policy perspective, the goals are to ensure that broadband deployment is timely and contributes to the nation's economic growth, that industry competes fairly, and that affordable and high-quality service is provided to all sectors and geographical locations of American society. The federal government — through Congress and the Federal Communications Commission (FCC) — is seeking to ensure fair competition among the players so that broadband will be available and affordable in a timely manner to all Americans who want it.

Chapter 3 - The "digital divide" is a term that has been used to characterize a gap between "information haves and have-nots," or in other words, between those Americans who use or have access to telecommunications technologies (e.g., telephones, computers, the Internet) and those who do not. One important subset of the digital divide debate concerns high-speed Internet access, also known as *broadband*. Broadband is provided by a series of technologies (e.g., cable, telephone wire, fiber, satellite, wireless) that give users the ability to send and receive data at volumes and speeds far greater than current "dial-up" Internet access over traditional telephone lines.

Broadband technologies are currently being deployed primarily by the private sector throughout the United States. While the numbers of new broadband subscribers continue to grow, studies and data suggest that the rate of broadband deployment in urban and high income areas may be outpacing deployment in rural and low-income areas. Some policymakers, believing that disparities in broadband access across American society could have adverse economic and social consequences on those left behind, assert that the federal government should play a more active role to avoid a "digital divide" in broadband access. One approach is for the federal government to provide financial assistance to support broadband deployment in underserved areas. Others, however, believe that federal assistance for broadband deployment is not appropriate. Some opponents question the reality of the "digital divide," and argue that federal intervention in the broadband marketplace would be premature and, in some cases, counterproductive.

Legislation introduced in the 1 10th Congress seeks to provide federal financial assistance for broadband deployment in the form of grants, loans, subsidies, and tax credits. Of particular note is the reauthorization and reform of the Rural Utilities Service broadband loan program, which was enacted as part of the 2008 farm bill (P.L. 110-234). Legislation to reform universal service — which could impact the amount of financial assistance available for broadband deployment in rural areas — has been introduced (H.R. 42, H.R. 2054, H.R. 3428, H.R. 6320, H.R. 6356, S. 101, S. 711), as well as other broadband legislation (H.R. 1818, H.R. 2035, H.R. 2174, H.R. 2272, H.R. 2419, H.R. 2569, H.R. 2720, H.R. 2764, H.R. 2953, H.R. 3246, H.R. 3281, H.R. 3627, H.R. 3893, H.R. 3919, H.R. 5682, S. 541, S. 761, S. 1032, S. 1190, S. 1264, S. 1439, S. 1492, S. 2242, S. 2302).

In assessing such legislation, several policy issues arise. For example, is the current status of broadband deployment data an adequate basis on which to base policy decisions? Is federal assistance premature, or do the risks of delaying assistance to underserved areas outweigh the benefits of avoiding federal intervention in the marketplace? And finally, if one assumes that governmental action is necessary to spur broadband deployment in underserved areas, which specific approaches, either separately or in combination, are likely to be most effective?

Chapter 4 - Given the large potential impact broadband access to the Internet may have on the economic development of rural America, concern has been raised over a "digital divide" between rural and urban or suburban areas with respect to broadband deployment. While there are many examples of rural communities with state of the art telecommunications facilities, recent surveys and studies have indicated that, in general, rural areas tend to lag behind urban and suburban areas in broadband deployment.

Citing the lagging deployment of broadband in many rural areas, Congress and the Administration acted in 2001 and 2002 to initiate pilot broadband loan and grant programs within the Rural Utilities Service (RUS) at the U.S. Department of Agriculture (USDA). Subsequently, Section 6103 of the Farm Security and Rural Investment Act of 2002 (P.L. 107-17 1) amended the Rural Electrification Act of 1936 to authorize a loan and loan guarantee program to provide funds for the costs of the construction, improvement, and acquisition of facilities and equipment for broadband service in eligible rural communities. Currently, RUS/USDA houses the only two federal assistance programs *exclusively* dedicated to financing broadband deployment: the Rural Broadband Access Loan and Loan Guarantee Program and the Community Connect Grant Program.

The 11 0th Congress considered reauthorization and modification of the program as part of the farm bill. H.R. 2419/H.R. 6124, the Food, Conservation, and Energy Act of 2008 became law on June 18, 2008 (P.L. 110-246). Title VI (Rural Development) contains authorizing language for the broadband loan program. Additionally, H.R. 2035, H.R. 2174, H.R. 2569, H.R. 2953, S. 541, S. 1032, and S. 1439 were introduced to address some of the reauthorization issues related to the RUS broadband loan program.

Meanwhile, on May 11, 2007, RUS released a Proposed Rule seeking to revise the broadband loan program rules and regulations. Some key issues pertinent to a consideration of the RUS broadband programs include restrictions on applicant eligibility, how "rural" is defined with respect to eligible rural communities, how to address assistance to areas with preexisting broadband service, technological neutrality, funding levels and mechanisms, and the appropriateness of federal assistance. The final rule will reflect language in the farm bill statute (P.L. 110-246). Ultimately, modification of rules, regulations, or criteria associated with the RUS broadband program will likely result in "winners and losers" in terms of which companies, communities, regions of the country, and technologies are eligible or more likely to receive broadband loans and grants.

Chapter 5 - About 30 million American households purchase, or have adopted, broadband service, but it is difficult to assess the extent of gaps in the availability of broadband in local markets. Using a survey of American households, we found that 28 percent—or about 30 million—subscribed to broadband service in 2005. In addition, 30 percent of surveyed households subscribed to a dial-up Internet service, and 41 percent did not access the Internet from their home. Among households subscribing to broadband service, we found roughly an equal share taking cable modem and digital subscriber line (DSL) service, the two primary broadband services at this time. Households in rural areas were less likely to subscribe to broadband service, compared with households in urban and suburban areas. On a semiannual basis, FCC conducts an extensive data collection effort using its Form 477 to assess the availability of advanced telecommunications service in the United States. As of July 2005, FCC has found that 99 percent of Americans live in the 95 percent of zip codes that have at least one broadband provider reporting to be serving at least one subscriber. These data clearly indicate that deployment of broadband networks has been extensive. However, for its zip-code level data, FCC collects data based on where *subscribers* are served, not where providers have deployed broadband infrastructure. Based on our analysis is appears that these data may not provide a highly accurate depiction of deployment of broadband infrastructures for residential service in some areas.

Chapter 6 - Today's hearing will focus on the challenges for providing reliable, affordable broadband access of the rural small businesses. Experience has shown that broadband can bring economic revitalization to small towns by creating clusters of small businesses. Broadband service can also help farmers and farm-related businesses control costs and optimize production. This technology can provide real-time access to whether reports, fertilization guidance, and livestock tracking.

In: Maximizing Broadband Services to Rural Communities ISBN 978-1-60692-846-2
Editors: V.B. Haynesworth, A.R. Harrison, pp.1-10

Chapter 1

USDA RURAL DEVELOPMENT: BRINGING BROADBAND TO RURAL AMERICA*

United States Department of Agriculture

BROADBAND IS A HIGH PRIORITY

A USDA Rural Development is committed to fulfilling the President and Congress' goal of bringing affordable broadband to all rural Americans. This is a high priority. Broadband is a transformative technology. It allows rural communities to dramatically enhance the quality of health care and education. It gives every rural business access to regional, national, and international markets. It reduces barriers of time and distance, levels the playing field, and makes rural communities better places to live, work, and raise a family.

Rural America is diverse and so is the broadband program. USDA Rural Development tailors solutions to the needs of individual communities. We have financed technologies including fiber to the home, DSL, wireless, broadband over power lines, and hybrid fiber/ coax systems. We are eager to work with startup companies as well as established providers and stand ready to consider innovative financing arrangements. Our mission is to increase

* Excerpted from Rural Broadband issued by USDA Rural Development, dated May, 2007.

economic opportunity and improve the quality of life in rural America. Broadband is essential to both of these goals, and we will not rest until the job is done.

Thomas Dorr
Under Secretary

RURAL BROADBAND

As an agency of the U.S. Department of Agriculture, Rural Development is focused on improving the quality of life for rural Americans. In essence, Rural Development provides individuals, communities, and businesses with the financial assistance needed to address the unique needs of rural America. It offers programs in housing, utilities, business development, and community facilities/public safety. By improving the quality of life in rural America, it no longer is a sacrifice for the next generation to return home and raise their families in a safe and comforting environment.

Rural Development provides increased economic opportunities so people who choose to live in small towns can compete on a global level.

PROGRAM HISTORY

USDA Rural Development's involvement with broadband began in 2000 with a broadband pilot program, established with the help of a $100 million loan program authorized by the President of the United States and Congress. The program provided loan funds to rural communities with up to 20,000 residents. During the first year, Rural Development approved 12 loans for the entire $100 million. In addition, an extra $80 million was provided by the Agriculture Appropriations Act of 2002, and a total of 16 loans were approved during the second year of the pilot program.

In total, the pilot program delivered 28 loans with $180 million based in 20 States. The pilot program was successful because it allowed the agency to quickly implement a program and get an idea of what the needs and interests were for broadband.

The broadband loans serve a total 1,263 communities of 582,000 household subscribers.

The program has, and will continue, to facilitate deployment of broadband in rural areas. Currently, 40 percent of the projects financed under this program are successfully completed. However, the pilot approach did not include the risk-mitigating rules that are in place today with regard to equity and loan security. As a result, approximately 30 percent of the pilot loans defaulted. Rural Development continues to work with these borrowers to negotiate financial arrangements to repay the loans. Rural Development used the experience as a learning tool by applying those valuable lessons learned to the Broadband Loan Program that exists today.

CURRENT PROGRAM

The current USDA Rural Development Broadband Program was established by the 2002 Farm Bill. The program is designed to facilitate the delivery of broadband services to rural America, ensuring that the organization is making sound lending decisions and positively investing taxpayer dollars.

Since its inception, the program has approved 70 loans in 40 states, totaling over $1.22 billion. The broadband loans serve 1,263 communities with a total of 582,000 household subscribers. Approximately 40 percent of these communities were unserved at the time of the loan approval, and an additional 15 percent had only one provider.

Moreover, the current portfolio of active projects serves 987 communities with 38 loans in progress, 15 in initial start-up phase, and 5 loans rescinded. There are eight completed projects in Kansas, Louisiana, Michigan, North Dakota, Nebraska, South Dakota, Texas, and Washington.

The program is committed to bringing solutions to rural America and works hard to ensure that approved loans are likely to result in new service to rural residents and businesses. As of April 2007, the program received and reviewed nearly 200 applications. More than 100 were returned because applicants were unable to generate the minimum equity required to support the project and/or had incomplete or unsubstantiated business plans.

Over time, the number of returned applications has decreased as USDA Rural Development provided tools and assistance to loan applicants.

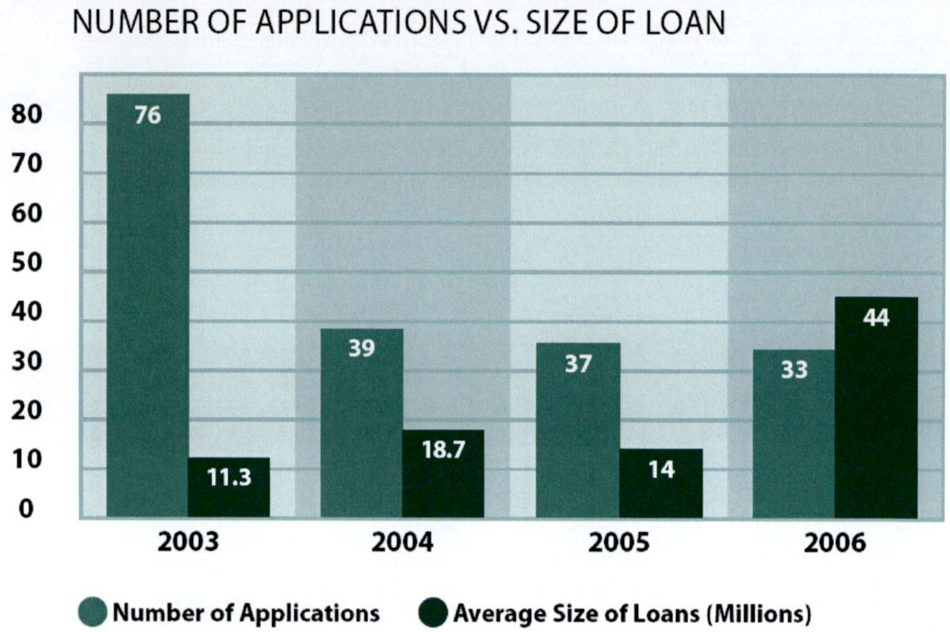

Demand to participate in the program is high, with pending fund requests currently at $981 million. Compared with previous years, fewer applications are being filed, but the requests that come in are for multicommunity or larger regional projects. Therefore, the average size of loans requested is increasing.

SUCCESS ACROSS RURAL AMERICA

The following are two examples of how broadband positively affected rural communities in Michigan and Kansas. Broadband technology brought much needed growth and opportunity for businesses, residents, and students.

Air Advantage

February 2003

Two Broadband Loans – $1,528,000

Two Community Connect Grants – $433,708

Air Advantage, founded in 2002, provides high-speed wireless Internet service to Michigan's "thumb" area. Scott zimmer, president of Air Advantage, worked with schools and saw the need for students to have broadband access. zimmer investigated how to make it happen. USDA Rural Development came to the rescue and assisted Air Advantage with two Community Connect Grants and two Broadband Loans. Air Advantage now serves 3,500 customers in four counties in Michigan, with another 1,000 customers projected under the second Broadband Loan.

The 300 miles of wireless backbone and 2,500 square miles of coverage have improved the lives of many rural Michigan residents. One facility that benefited from the broadband technology is the Columbia Township Library. What was once a one-room schoolhouse converted into a library in 1965 is now a brand new library that offers a host of services to local residents. Since 2005, the library computer lab has averaged more than 350 users (out of 605 area residents) per month. In addition, the increased use of the library and computer lab has given the facility an opportunity to offer community programs, such as a young reader program and community outreach programs.

Another successful result of the broadband technology involves the implementation of a distance learning program with seven school districts. This project has connected the schools via wireless broadband technologies and has allowed them to share teacher resources, giving students more educational opportunities. The distance learning program would not be available without the equipment obtained through the Distance Learning and Telemedicine Grant Program from USDA Rural Development. Some key offerings that have come about because of this program are foreign language and Advanced Placement courses that were not offered to students previously. Distance learning makes these courses available to all Huron County students because one teacher in one district can provide instruction to students in multiple districts.

"I could write a book about all of the wonderful things Rural Development has done for our company and our communities."

– Scott Zimmer

TECHNOLOGIES/TYPES OF ORGANIZATIONS

As the Rural Development Broadband Program is "technology neutral," a variety of technologies have been deployed by its borrowers. In addition, USDA Rural Development makes loans to diverse organizations.

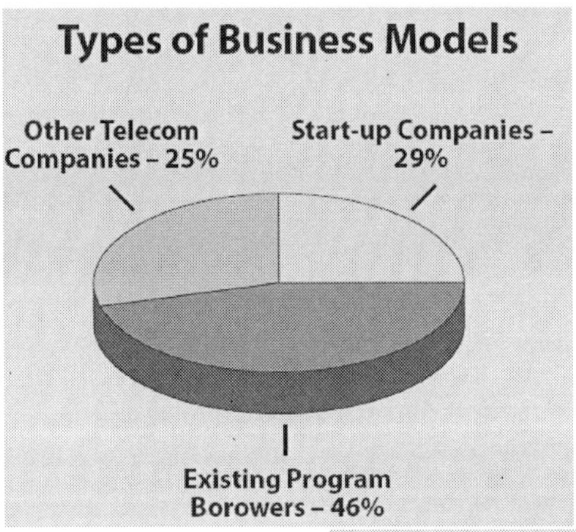

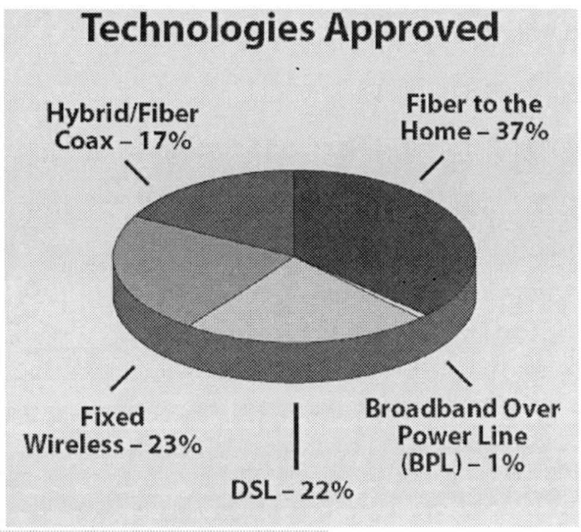

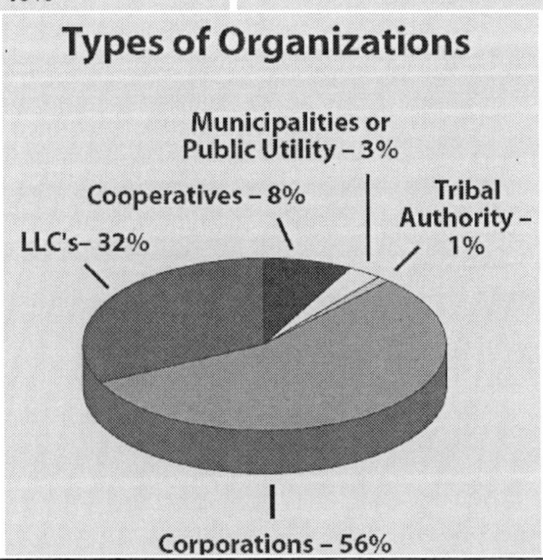

There are more than 50 students in Huron County enrolled in college-level courses while attending high school. These dual enrollment options are available because of the distance learning equipment. This program prepares the student for college, eliminates the need for driving up to 60 miles for this instruction, and permits the student to participate in these courses.

Thanks to Air Advantage and USDA Rural Development, the citizens of rural Michigan now have extensive opportunities to grow and thrive in their community.

Nex-Tech

November 2004

Two Broadband Loans – $6,574,000

Nex-Tech, a subsidiary of Rural Telephone, a leader in the telecommunications industry, has brought broadband, video, and voice services to towns throughout rural Kansas and has changed the future of these towns indefinitely. The Broadband Loans through USDA Rural Development allowed Nex-Tech to bring high-speed data connections to communities that had little or no connectivity. As a result of these long- and short-term economic benefits, these towns flourished.

Broadband technology allowed businesses to grow and expand, which provided a catalyst to hire and train more workers from the area. Prior to Nex-Tech, these businesses were using a slow dial-up connection that was not an efficient means of meeting business demands. The future development of these businesses relied on broadband becoming available, and some companies were even considering relocating if they did not get access.

For example, Osborne Industries, a company that has been in operation for more than three decades, was considering leaving the town of Osborne due to the lack of broadband availability. Since Nex-Tech's overbuild, Osborne Industries has expanded its product line and is able to reach markets throughout the United States, Canada, Europe, and Asia. Transitioning to electronic files has made ordering, servicing, and accounting more efficient. The company supports 112 employees.

Fiber optics and broadband Internet have played an important role in attracting and retaining residents and businesses to the Phillips County area. Since the new technology, there has been a slight increase in school enrollment across the three school districts in the county. The population is estimated to be increasing with new home construction on the rise and a variety of businesses that are expanding and relocating to the area. For instance, a $54 million ethanol plant that employs 34 people was built in Phillipsburg after broadband became available.

Numerous improvements have been made with regard to high-speed Internet connections, allowing businesses to operate more efficiently and allowing executives to work from home through a virtual private network (vPN). Nex-Tech and USDA Rural Development understand the importance of community development and have a long-term vision for rural communities by continuing to grow and expand their advanced data services to reach more residents.

PROGRAM IMPROVEMENTS

Although the Broadband Program is working as Congress intended with $1.22 billion invested to date, changes are being made to improve results.

Under the current Broadband Program, USDA Rural Development has:

- Reduced processing times from an average of 331 days to 181 days.
- Reallocated 11 employees from other program areas to ensure pending applications are reviewed sooner.
- Improved access to program information by:
  - Creating a new USDA Rural Development Web link on the Federal Communications Commission (FCC) rural homepage. – Redesigning the program Web site for easier navigation and improved usability.
  - Posting weekly information about pending and approved broadband loans on the program Web site.
  - Posting a Traditional Borrower listing that allows applicants to identify areas that may be ineligible for funding.

In addition, the Program is reviewing its processes and procedures to identify improvement or streamlining opportunities.

STEPS TOWARD CHANGE

USDA Rural Development is committed to improving the Broadband Loan Program's effectiveness. To better meet the needs of rural America, Rural Development has proposed a series of improvements to the Broadband Program. The proposed rule modifications are designed to:

- Promote deployment to rural areas with little or no service.
- Limit funding in urban areas where a significant share of the market is served by incumbent providers.
- Ensure that residents in funded areas get broadband access more quickly.
- Clarify and streamline the equity and marketing survey requirements.
- Increase the transparency of the application process.
- Promote a better understanding of all application requirements.
- Ensure that projects funded are keeping pace with increasing demand for bandwidth.

Advocate for Rural America

USDA Rural Development is the advocate for rural America and has been for many decades. Its Broadband Program supports that effort with its vision for rural communities — to have access to affordable, reliable, advanced communications services to provide a healthy, safe, and prosperous place to live and work. With the continued support and dedicated efforts of USDA Rural Development employees along with the residents of the rural communities they assist, this vision is becoming a reality.

Over the years, USDA Rural Development has established expertise in the broadband arena and employed a team-focused approach to improving life in rural communities. Currently, the Broadband Program is administered by a staff of engineers and loan analysts who understand the industry and its unique challenges. Existing USDA Rural Development distribution systems also support the broadband effort with field representatives and offices in all 50 states. The program staff provides financial and engineering support to applicants prior to and after filing, while state offices, field representatives, and headquarters staff conduct broadband workshops frequently to inform potential applicants and community leaders about the program and the benefits of broadband. To date, the program has conducted 23 broadband workshops in addition to many other outreach events.

Looking Forward

USDA Rural Development is committed to the future of rural communities, ensuring they are great places in which to live, work, and raise families.

It has made great progress in bringing broadband to rural America, but there is much more to do. The organization will continue to refine the Broadband Program so that more rural residents and businesses can receive the economic and social benefits that broadband creates in their communities. In addition, it will continue to rely on the close support and dedicated efforts of the residents within those rural communities to help facilitate those investments. With this partnership, broadband will be an integral part of life in rural communities across the country, providing the economic benefits of the city with the high quality of life typically found in the heart of rural America.

In: Maximizing Broadband Services to Rural Communities ISBN 978-1-60692-846-2
Editors: V.B. Haynesworth, A.R. Harrison, pp. 11-34

Chapter 2

BROADBAND INTERNET REGULATION AND ACCESS: BACKGROUND AND ISSUES*

Angele A. Gilroy and Lennard G. Kruger

ABSTRACT

Broadband or high-speed Internet access is provided by a series of technologies that give users the ability to send and receive data at volumes and speeds far greater than current Internet access over traditional telephone lines. In addition to offering speed, broadband access provides a continuous, "always on" connection and the ability to both receive (download) and transmit (upload) data at high speeds. Broadband access, along with the content and services it might enable, has the potential to transform the Internet: both what it offers and how it is used. It is possible that many of the future applications that will best exploit the technological capabilities of broadband have yet to be developed. There are multiple transmission media or technologies that can be used to provide broadband access. These include cable; an enhanced telephone service called digital subscriber line (DSL); fiber-tothe-home (FTTH); satellite, mobile, and fixed wireless (including "wi-fi" and "Wi-Max"); broadband over powerlines (BPL); and others.

From a public policy perspective, the goals are to ensure that broadband deployment is timely and contributes to the nation's economic growth, that industry competes fairly, and that affordable and high-quality service is provided to all sectors and geographical locations of American society. The federal government — through Congress and the Federal Communications Commission (FCC) — is seeking to ensure fair competition among the players so that broadband will be available and affordable in a timely manner to all Americans who want it.

Some areas of the nation — particularly rural and low-income communities — continue to lack full access to high-speed broadband Internet service. In order to address this problem, the 110th Congress is examining a wide range of issues including the scope and effect of federal broadband financial assistance programs (including universal service and the broadband programs at the U.S. Department of Agriculture's Rural Utilities Service), and

* Excerpted from CRS Report for Contgress, Order Code RL33542, dated August 25, 2008.

the impact of telecommunications regulation and new technologies on broadband deployment. To date, legislative measures to address the reform and expansion of scope of the universal service fund (S. 101, S. 609, S. 711, H.R. 42, H.R. 278, H.R. 2054 H.R. 6320, H.R. 6356), net neutrality (S. 215. H.R. 5353, H.R. 5994), and broadband financial assistance and data collection (H.R. 1818, H.R. 2035, H.R. 2174, H.R. 2272, H.R. 2419, H.R. 2569, H.R. 2764, H.R. 2953, H.R. 3246, H.R. 3281, H.R. 3428, H.R. 3627, H.R. 3893, H.R. 3919, H.R. 5682, S. 541, S. 761, S. 1032, S. 1190, S. 1264, S. 1439, S. 1492, S. 2242) have been introduced. One facet of the debate over broadband services focuses on whether present laws and subsequent regulatory policies are needed to ensure the development of competition and its subsequent consumer benefits, or conversely, whether such laws and regulations are overly burdensome and discourage investment in and deployment of broadband services.

WHAT IS BROADBAND AND WHY IS IT IMPORTANT?

Broadband or high-speed Internet access is provided by a series of technologies that give users the ability to send and receive data at volumes and speeds far greater than current Internet access over traditional telephone lines. Currently, a number of telecommunications companies are developing, installing, and marketing specific technologies and services to provide broadband access to the home. Meanwhile, the federal government — through Congress and the Federal Communications Commission (FCC) — is seeking to ensure fair competition among the players so that broadband will be available and affordable in a timely manner to all Americans who want it.

Traditionally, Internet users have accessed the Internet through the same telephone line that can be used for traditional voice communication. A personal computer equipped with a modem is used to hook into an Internet dial-up connection provided (for a fee) by an Internet service provider (ISP) of choice. The modem converts analog signals (voice) into digital signals that enable the transmission of "bits" of data.

The faster the data transmission rate, the easier one can download files, hop from Web page to Web page, or view video. The highest speed modem used with a traditional telephone line, known as a 56K modem, offers a maximum data transmission rate of about 45,000 bits per second (bps). However, as the content on the World Wide Web becomes more sophisticated, the limitations of relatively low data transmission rates (called "narrowband") such as 56K become apparent. For example, using a 56K modem connection to download a 10-minute video or a large software file can be a lengthy and frustrating exercise. By using a broadband highspeed Internet connection, with data transmission rates many times faster than a 56K modem, users can view video, make telephone calls, or download software and other data-rich files in a matter of seconds. In addition to offering speed, broadband access provides a continuous "always on" connection (no need to "dial-up") and a "two- way" capability — that is, the ability to both receive (download) and transmit (upload) data at high speeds.

Broadband access, along with the content and services it might enable, has the potential to transform the Internet — both what it offers and how it is used. For example, a two-way high speed connection could be used for interactive applications such as online classrooms, showrooms, or health clinics, where teacher and student (or customer and salesperson, doctor and patient) can see and hear each other through their computers. An "always on" connection could be used to monitor home security, home automation, or even patient health remotely through the Web. The high speed and high volume that broadband offers could also be used for bundled service where, for example, cable television, video on demand, voice, data, and other services are all offered over a single line. In truth, it is possible that many of the applications that will best exploit the technological capabilities of broadband, while also capturing the imagination of consumers, have yet to be developed.

Broadband Technologies

There are multiple transmission media or technologies that can be used to provide broadband access. These include cable modem, an enhanced telephone service called digital subscriber line (DSL), satellite technology, fiber, mobile or fixed wireless technologies, and others. Cable and DSL are currently the most widely used technologies for providing broadband access. Both require the modification of an existing physical infrastructure that is already connected to the home (i.e., cable television and telephone lines). Each technology has its respective advantages and disadvantages, and competes with each other based on performance, price, quality of service, geography, user friendliness, and other factors. The following sections summarize cable, DSL, and other broadband technologies.

Cable

The same cable network that currently provides television service to consumers is being modified to provide broadband access. Because cable networks are shared by users, access speeds can decrease during peak usage hours, when bandwidth is being shared by many customers at the same time. Network sharing has also led to security concerns and fears that hackers might be able to eavesdrop on a neighbor's Internet connection. The cable industry is developing "next generation" technology which will significantly extend downloading and uploading speeds.

Digital Subscriber Line (DSL)

DSL is a modem technology that converts existing copper telephone lines into two-way high speed data conduits. Speeds can depend on the condition of the telephone wire and the distance between the home and the telephone company's central office (i.e., the building that houses telephone switching equipment). Because DSL uses frequencies much higher than those used for voice communication, both voice and data can be sent over the same telephone line. Thus, customers can talk on their telephone while they are online, and voice service will continue even if the DSL service goes down. Like cable broadband technology, a DSL line is "always on" with no dial-up required. Unlike cable, however, DSL has the advantage of being unshared between the customer and the central office. Thus, data transmission speeds will not necessarily decrease during periods of heavy local Internet use. A disadvantage relative to cable is that DSL deployment is constrained by the distance between the subscriber and the central office. DSL technology over a copper wire only works within 18,000 feet (about three miles) of a central office facility. However, DSL providers are deploying technology to further increase deployment range. One option is to install “remote terminals” which can serve areas farther than three miles from the central office.

Wireless

Mobile or fixed wireless systems transmit data over the airwaves from towers or antennas to a receiver. Mobile wireless broadband services (also referred to as third generation or “3G”) allow consumers to get broadband access over cell phones, PDAs, or wireless modem cards connected to a laptop.[1] The FCC has auctioned frequencies currently occupied by broadcast channels 52-69. These and other frequencies in the 700 MHZ band are possible candidates for wireless broadband applications. A number of wireless technologies, corresponding to different parts of the electromagnetic spectrum, also have potential. These include the upperbands (above 24GHz), the lowerbands (multipoint distribution service or MDS, below 3 GHz), broadband personal communications services (PCS), wireless communications service (2.3 GHz), and unlicenced spectrum. Unlicensed spectrum is being increasingly used to provide high-speed short-distance wireless access (popularly called “wi-fi”) to local area networks, particularly in urban areas where wired broadband connections already exist. A new and developing wireless broadband technology (called “WiMax”) has the capability to transmit signals over much larger areas.

Fiber

Another broadband technology is optical fiber to the home (FTTH). Optical fiber cable, already used by businesses as high speed links for long distance voice and data traffic, has tremendous data capacity, with transmission speeds dramatically higher than what is offered by cable modem or DSL broadband technology. While the high cost of installing optical fiber in or near users’ homes has been a major barrier to the deployment of FTTH, both Verizon and AT&T (formerly SBC) are rolling out fiber-based architectures that will offer consumers voice, video, and highspeed data (sometimes referred to as a “triple play”). Some public utilities are also exploring or beginning to offer broadband access via fiber inside their existing conduits. Additionally, some companies are investigating the feasibility of transmitting data over power lines, which are already ubiquitous in people’s homes.[2]

Satellite

Satellite broadband Internet service is currently being offered by three providers: Hughes Network Systems (DirecWay), Starband (Spacenet Inc.) and WildBlue. Like cable, satellite is a shared medium, meaning that privacy may be compromised and performance speeds may vary depending upon the volume of simultaneous use. Another disadvantage of Internet - over-satellite is its susceptibility to disruption in bad weather. On the other hand, the big advantage of satellite is its universal availability. Whereas cable or DSL is not available to some parts of the United States, satellite connections can be accessed by anyone with a satellite dish facing the southern sky. This makes satellite Internet access a possible solution for rural or remote areas not served by other technologies.

STATUS OF BROADBAND DEPLOYMENT

According to the latest FCC data on the deployment of high-speed Internet connections (released March 2008), as of June 30, 2007, there were 100.9 million high speed lines connecting homes and businesses to the Internet in the United States, a growth rate of 22% during the first half of 2007. Of the 100.9 million high speed lines reported by the FCC, 65.9 million serve residential users.[3] While the broadband *adoption* rate stands at roughly 58% of U.S. households,[4] broadband *availability* is much higher. As of June 30, 2007, the FCC found at least one high-speed subscriber in 99% of all zip codes in the United States. The FCC estimates that "roughly 20 percent of consumers with access to advanced telecommunications capability do subscribe to such services." According to the FCC, possible reasons for the gap between broadband availability and subscribership include the lack of computers in some homes, price of broadband service, lack of content, and the availability of broadband at work.[5]

According to the International Telecommunications Union, the U.S. ranks 24th worldwide in broadband penetration (subscriptions per 100 inhabitants in 2007).[6] Data from the Organization for Economic Cooperation and Development (OECD) found the U.S. ranking 15th among OECD nations in broadband access per 100 inhabitants as of December 2007.[7] By contrast, in 2001 an OECD study found the U.S. ranking 4th in broadband subscribership per 100 inhabitants (after Korea, Sweden, and Canada).[8] While many argue that the U.S. declining performance in international broadband rankings is a cause for concern,[9] others — including the Administration — maintain that the OECD and ITU data undercount U.S. broadband deployment,[10] and that cross-country broadband deployment comparisons are not necessarily meaningful and inherently problematic.[11] Finally, an issue related to international broadband rankings is the extent to which broadband speeds and prices differ between the U.S. and the rest of the world.[12]

ACCESS TO BROADBAND AND THE "DIGITAL DIVIDE"[13]

While the number of new broadband subscribers continues to grow, the rate of broadband deployment in urban and high income areas appears to be outpacing deployment in rural and low-income areas. According to the latest FCC data on the deployment of high-speed Internet connections (released March 2008), high-speed subscribers were reported in 99% of the most densely populated zip codes, as opposed to 91% of zip codes with the lowest population densities. Similarly, for zip codes ranked by median family income, high-speed subscribers were reported present in 99% of the top one-tenth of zip codes, as compared to 92% of the bottom one- tenth of zip codes.[14]

Similarly, 2008 data from the Pew Internet & American Life Project indicate that while broadband adoption is growing in urban, suburban, and rural areas, broadband users make up larger percentages of urban and suburban users than rural users. Pew found that the percentage of all U.S. adults with broadband at home is 60% for suburban areas, 57% for urban areas, and 38% for rural areas.[15]

Some policymakers assert that disparities in broadband access across American society could have adverse consequences on those left behind. Many believe that advanced Internet applications — voice over the Internet protocol (VoIP) or high quality video, for example — and the resulting ability for businesses and consumers to engage in e-commerce, may increasingly depend on high speed broadband connections to the Internet. Thus, some say, communities and individuals without access to broadband could be at risk to the extent that e-commerce becomes a critical factor in determining future economic development and prosperity.

FCC Activities

The Telecommunications Act of 1996 (P.L. 104-104) addressed the issue of whether the federal government should intervene to prevent a "digital divide" in broadband access. Section 706 requires the FCC to determine whether "advanced telecommunications capability [i.e., broadband or high-speed access] is being deployed to all Americans in a reasonable and timely fashion." If this is not the case, the act directs the FCC to "take immediate action to accelerate deployment of such capability by removing barriers to infrastructure investment and by promoting competition in the telecommunications market."

On September 9, 2004, the FCC adopted and released its *Fourth Report* pursuant to Section 706. Like the previous three reports, the FCC concluded that "the overall goal of section 706 is being met, and that advanced telecommunications capability is indeed being deployed on a reasonable and timely basis to all Americans."[16] While the FCC is currently implementing or actively considering some regulatory activities related to broadband,[17] no major regulatory intervention pursuant to Section 706 of the Telecommunications Act of 1996 has been deemed necessary by the FCC at this time.

The FCC noted the future promise of emerging multiple advanced broadband networks which can complement one another:

> For example, in urban and suburban areas, wireless broadband services may "fill in the gaps" in wireline broadband coverage, while wireless and satellite services may bring high-speed broadband to remote areas where wireline deployment may be costly. Having multiple advanced networks will also promote competition in price, features, and quality-of-service among broadband-access providers.[18]

Two FCC Commissioners (Michael Copps and Jonathan Adelstein) dissented from the *Fourth Report* conclusion that broadband deployment is reasonable and timely. They argued that the relatively poor world ranking of United States broadband penetration indicates that deployment is insufficient, that the FCC's continuing definition of broadband as 200 kilobits per second is outdated and is not comparable to the much higher speeds available to consumers in other countries, and that the use of zip code data (measuring the presence of at least one broadband subscriber within a zip code area) does not sufficiently characterize the availability of broadband across geographic areas.[19]

The Government Accountability Office (GAO) has also cited concerns about the FCC's zip code level data. Of particular concern is that the FCC will report broadband service in a zip code even if a company reports service to only one subscriber, which in turn can lead to some observers overstating of broadband deployment. According to GAO, "the data may not provide a highly accurate depiction of local deployment of broadband infrastructures for residential service, especially in rural areas."[20]

On March 19, 2008, the FCC adopted an Order that substantially expands its broadband data collection capability. Specifically, the Order expands the number of broadband reporting speed tiers to capture more information about upload and download speeds offered in the marketplace, requires broadband providers to report numbers of broadband subscribers by census tract, and improves the accuracy of information collected on mobile wireless broadband deployment. Additionally, in a Further Notice of Proposed Rulemaking, the FCC is seeking comment on broadband service pricing and availability.[21]

Also on March 19, 2008, the FCC adopted the *Fifth Report* to Congress on broadband deployment under Section 706 of the Telecommunications Act of 1996. As did previous reports, the Fifth Report found that broadband services are currently being deployed to all Americans in a reasonable and timely fashion. Commissioners Copps and Adelstein again dissented, citing flawed data collection methodologies, lagging U.S. broadband penetration internationally, and the lack of a comprehensive U.S. broadband strategy.

Administration Activities

The National Telecommunications and Information Administration (NTIA) at the Department of Commerce (DOC) has been tasked with developing the Bush Administration's broadband policy.[22] Statements from Administration officials indicated that much of the policy would focus on removing regulatory roadblocks to investment in broadband deployment.[23] On June 13, 2002, in a speech at the 21st Century High Tech Forum, President Bush declared that the nation must be aggressive about the expansion of broadband, and cited ongoing activities at the FCC as important in eliminating hurdles and barriers to get broadband implemented. President Bush made similar remarks citing the economic importance of broadband deployment at the August 13, 2002 economic forum in Waco, Texas. Subsequently, a more formal Administration broadband policy was unveiled in March and April of 2004. On March 26, 2004, President Bush endorsed the goal of universal broadband access by 2007.[24] Then on April 26, 2004, President Bush announced a broadband initiative which advocates permanently prohibiting all broadband taxes, making spectrum available for wireless broadband, creating technical standards for broadband over power lines, and simplifying rights-of-way processes on federal lands for broadband providers.[25]

On January 31, 2008, NTIA released a report, entitled, *Networked Nation: Broadband in America, 2007.*[26] According to NTIA, the report shows "that the Administration's technology, regulatory, and fiscal policies have stimulated innovation and competition, and encouraged investment in the U.S. broadband market contributing to significantly increased accessibility of broadband services."[27]

Enacted Legislation

Some policymakers in Congress have asserted that the federal government should play a more active role to avoid a "digital divide" in broadband access, and that legislation is necessary to ensure fair competition and timely broadband deployment. The Farm Security and Rural Investment Act of 2002 — signed into law on May 13, 2002 as P.L. 107-171 — contained a provision (Section 6103) authorizing the Secretary of Agriculture to make loans and loan guarantees to eligible entities for facilities and equipment providing broadband service in rural communities. The 110th Congress has reauthorized and reformed the RUS broadband loan program as part of the 2008 farm bill (P.L. 110-234).[28]

Congress has also enacted legislation intended to make radiofrequency spectrum available for wireless broadband applications. For example, the 108th Congress enacted The Commercial Spectrum Enhancement Act (Title II of P.L. 108-494), which seeks to make more spectrum available for wireless broadband and other services by facilitating the reallocation of spectrum from government to commercial users. In the 109th Congress, the Title III of the Deficit Reduction Act of 2005 (P.L. 109-171) set a hard deadline for the digital television transition, thereby reclaiming analog television spectrum to be auctioned for commercial applications such as wireless broadband.

Regulation and Broadband: Convergence and the Changing Marketplace

Rapid technological advances and the resulting convergence of telecommunications providers and markets has prompted the reexamination of the existing telecommunications industry regulatory framework. The "Telecommunications Act of 1996," (P.L.104-104) redefined and recast the 1934 Communications Act to address the emergence of competition in what were previously considered to be monopolistic markets. Despite its relatively recent enactment, however, a consensus has been growing that the modifications brought about by the implementation of the 1996 Act are not sufficient to address the Nation's changing telecommunications environment. Technological changes such as the advancement of Internet technology to supply data, voice, and video as well as the growing convergence in the telecommunications sector, have, according to many policymakers, made it necessary to consider another "rewrite" or revision of the laws governing these markets.

The regulatory debate focuses on a number of issues including the extent to which existing regulations should be applied to traditional providers as they enter new markets where they do not hold market power, the extent to which existing regulations should be imposed on new entrants as they compete with traditional providers in the same markets, and the appropriate regulatory framework to be imposed on new and/or converging technologies that are not easily classified under the present framework.[29]

The regulatory treatment of broadband technologies continues to hold a major focus in the policy debate. A major facet of the debate centers on whether present laws and regulations are needed to ensure the development of competition and its subsequent consumer benefits, or, conversely, whether such laws and policies are overly burdensome and discourage needed investment and deployment of such services. What if any role regulators should play to ensure the Internet remains open to all, often referred to as "open access" requirements or "net neutrality," is also a major and contentious part of the dialogue.[30] In addition to the debate over economic regulation, concern over how and to what extent "social regulations" such as emergency 911 access, disability access, and law enforcement regulations, should be applied to new and converging technologies continues to be debated. The continued growth and expressed interest in municipal broadband networks has also focused debate on what the appropriate role of the government sector should be and whether it should be competing with the private sector.

How traditional policy goals, such as the advancement of universal service mandates, should be revised to accommodate the changing marketplace has also come under scrutiny. For example, issues such as who should receive and who should contribute to universal service funds and whether the definition of universal service objectives should be expanded to include new technologies such as broadband continue to be debated.[31]

ACTIVITIES IN THE 109TH CONGRESS

In the 1 09th Congress, debate over broadband policy primarily centered on H.R. 5252 — the Communications Opportunity, Promotion, and Enhancement Act (COPE) in the House, and the Advanced Telecommunications and Opportunity Reform Act (ATOR) in the Senate. H.R. 5252 addressed a number of issues, including the extent to which legacy regulations should be applied to traditional providers as they enter new markets, the extent to which legacy regulations should be imposed on new entrants as they compete with traditional providers in their markets, the treatment of new and converging technologies, and the emergence of municipal broadband networks and Internet access. H.R. 5252, as amended, passed (321-101) the House, was significantly amended and passed (15-7) by the Senate Commerce Committee, but did not reach the Senate floor for consideration.

H.R. 5252 (COPE)

House Commerce Committee Chairman Barton, on March 27, 2006, released a draft telecommunications reform proposal that was the subject of a Committee hearing on March 30, 2006. The then unnumbered measure, passed (27-4) the subcommittee, with amendment, on April 5, 2006, and passed (42-12) the full Committee with amendment, on April 26, 2006. The measure, titled "The Communications Opportunity, Promotion, and Enhancement Act of 2006" (COPE), was referred to the House Committee on Energy and Commerce and formally introduced as H.R. 5252. A sequential referral request, by House Judiciary Chairman Sensenbrenner, which was subsequently denied, delayed floor consideration. The House passed (321-101) an amended version of H.R. 5252 on June 8, 2006. In addition to a manager's amendment clarifying franchising provisions, five additional amendments were passed. The other amendments: established a complaint process to resolve fee disputes between a local franchise authority and a cable operator; increased the income discrimination penalty for a cable operator from $500,000 to $750,000; allowed a cable franchising authority to issue an order requiring compliance with FCC revised consumer protection rules; preserved

FCC authority to require VOIP providers to contribute to the federal universal service fund, when they connect directly or indirectly to the public switched network and compensate network owners for use of their network; and clarified that language in HR5252 giving the FCC the exclusive authority to adjudicate network neutrality does not remove antitrust authority over net neutrality complaints. Two amendments did not pass. The first, an amendment, sponsored by Representative Markey, to strengthen net neutrality provisions failed by a vote of 152-269. The second, to reduce, from 1 percent to 0.5 percent, the fee paid to local franchise authorities relating to PEG/iNet support by women-owned, small business and socially and economically disadvantaged firms was withdrawn.

H.R. 5252, as passed by the House, contained in its 6 titles, provisions that would establish a national cable franchising process; clarify the FCC's authority to enforce its network neutrality principles; address VoIP 911 interconnection and E911 requirements; and bar states from prohibiting municipalities from providing their own broadband networks. More specifically, Title I establishes a national process, through the FCC, for new entrants to offer pay TV services and opens it up to incumbent cable providers, once they face local competition. An operator of a national franchise is prohibited from discriminating in the provision of service to any group of residential subscribers based on the income of that group. National consumer protection rules are established with a local authority/FCC complaint procedure. Additional provisions in Title I preserve the local five percent franchise fee cap, preserve and support PEG channel and I-Nets or Institutional Networks (a one percent gross revenue fee is established to ensure financial support), and preserve rights-of-way requirements. The bill also contains provisions to assist small and rural carriers in the provision of video service by allowing video operators to share a headend transmission facility.

Title II clarifies the FCC's authority to enforce its August 2005 network neutrality principles in complaint proceedings, but prohibits the FCC from engaging in related-rulemaking. Fines up to $500,000 per violation are established and the FCC is required to resolve complaints within 90 days. The FCC is also directed to conduct and submit to the House Energy and Commerce and Senate Commerce Committees, within 180 days of enactment, a study, to evaluate ".... whether the objectives of the (FCC's) broadband policy statement and the principles incorporated therein are being achieved."

The remaining four titles dealt with a wide range of telecommunications issues. Title III of the bill contains provisions to establish 911 and E-9 11 requirements for VoIP services that connect to the public switched network and represent a replacement telephone service. Additional provisions provide access to the nation's 911 infrastructure and requires the FCC to appoint a 911 number administrator. Title IV contains provisions that bar states from prohibiting municipalities from providing their own broadband networks (that is telecommunications, information, or cable services), but also requires that they do not discriminate in favor of, or bestow any advantages to, such entities as compared with other providers of such services. The FCC is tasked with submitting within one year of enactment, a report to Congress, on the status of the provision of such services by municipalities. Titles V and VI contain provisions that ensue consumers can buy stand-alone broadband service; call for an FCC study to examine the possible interference associated with the deployment of broadband over power lines; and further the development of "seamless mobility."

S. 2686 (HR5252/ATOR)

The Senate Commerce Committee held a series of hearings on a wide range of telecommunications issues in preparation for developing comprehensive telecommunications legislation. Senate Commerce Committee Chairman Stevens introduced, on May 1, 2006, a comprehensive (135 page) telecommunications bill, S. 2686. The major provisions of that measure dealt with a wide range of topics, including universal service reform; streamlining of the video franchising process; requiring the FCC to report annually to Congress on the net neutrality issue; interoperability of public safety communications systems; interconnection; and municipal broadband ownership. The bill also contains a number of provisions relating to broadcast issues such as the digital television transition, the reinstating of the FCC's "broadcast flag" rules, access to sports programming, and use of unlicensed "white space." Additional provisions relating to protecting children from child pornography and amending the FCC's "sunshine rules" are also included.

Although Senator Inouye, the ranking minority member of the Committee, signed on as a bill co-sponsor, he stated that S. 2686 needed considerable amendment to gain his support. He circulated a draft proposal containing provisions addressing video franchising, Internet access, broadband deployment, and universal service, for consideration that addressed his concerns. The lack of a strong net neutrality provision was one of the issues he specifically singled out for attention. S. 2686 provisions relating to streamlining the video franchising process, universal service fund reform, and net neutrality were the major focus of Commerce Committee hearing held on May 18, and May 25,2006. The Commerce Committee issued a revised draft of the bill which was the subject of a hearing held on June 13, 2006.

After a lengthy and intense markup the Senate Commerce Committee approved (15-7) on June 28, 2006 the newly titled "Advanced Telecommunications and Opportunity Reform Act," which technically is an amended version in the nature of a substitute for H.R. 5252. In addition to a new bill name and number the three-day markup led to the approval of a significant manager's amendment containing a new title and 70 amendments resulting in the passage of a 200-plus page omnibus telecommunications measure. S. 2686, which was referred to as "the Senate Committee passed version of H.R. 5252," contains 11 titles covering a wide range of telecommunications issues including video franchise reform, net neutrality, universal service reform, municipal broadband, broadcast flag, the digital television transition, interoperability, the illegal transmission of child pornography, and FCC reform. The issue of net neutrality proved to be major point of contention during the markup. Despite the addition of a new title (Title IX) establishing an "Internet Consumer Bill of Rights" net neutrality advocates continued to press for a net neutrality non-discrimination provision. A nondiscrimination amendment offered during markup was defeated by an 11-11 vote. The lack of a cable franchise build-out provision, federal preemption of state authority over wireless services, as well as provisions added during markup to exempt, for three years, wireless providers from"new and discriminatory" taxes and make permanent the Internet tax moratorium also resulted in concern. While Senator Steven's continued to express confidence that the Senate version of H.R. 5252 would come to the floor for a vote, the 109th Congress ended without full Senate consideration of the measure.

Both the Senate and House Judiciary Committees also examined issues related to telecommunications reform. The House Judiciary's Telecommunications and Antitrust Task Force held a hearing on April 25, 2006, to examine competition issues relating to Internet access and "net neutrality." House Judiciary Committee Chairman Sensenbrenner and Representative Conyers, the ranking minority member, stated, in a letter sent to then House Speaker Hastert, that the Judiciary Committee had oversight over market conditions, consolidations and antitrust protections in the telecommunications sector, and asked for a sequential referral of H.R. 5252. That request was denied. However, Chairman Sensenbrenner, Representative Conyers and others introduced a bipartisan bill (H.R. 5417) focusing on Internet access from an antitrust perspective, that passed (20-13) the Judiciary Committee, with amendment, on May 25, 2006. A request to the House Rules Committee to have the bill considered as an amendment during House floor action on H.R. 5252 was denied. The Senate Judiciary Committee held a June 14, 2006 hearing to examine communications laws in the context of ensuring competition and innovation.

ACTIVITIES IN THE 110TH CONGRESS

In the 110th Congress, legislation has been introduced that would provide financial assistance for broadband deployment. Of particular note is the reauthorization and reform of the Rural Utilities Service (RUS) broadband loan program, which was enacted as part of the 2008 farm bill (P.L. 110-234).[32] P.L. 110- 234 also contains provisions establishing a National Center for Rural Telecommunications Assessment and requiring the FCC and RUS to formulate a comprehensive rural broadband strategy.

Legislation to reform universal service (H.R. 2054, H.R. 6320, H.R. 6356, S. 101, S. 711) which could have a significant impact on the amount of financial assistance available for broadband deployment in rural and underserved areas — has also been introduced. Additionally, Congress is considering broadband data bills (S. 1492 as reported by the Senate Commerce, Science, and Transportation Committee, and H.R. 3919 as passed by the House), net neutrality bills (H.R. 5353, H.R. 5994, S. 215), and municipal broadband bills (H.R. 3281 and S. 1853). The following provides a listing of broadband-related legislation introduced into the 110th Congress.

P.L. 110-69 (H.R. 2272)

America COMPETES Act. Authorizes the National Science Foundation (NSF) to provide grants for basic research in advanced information and communications technologies. Areas of research include affordable broadband access, including wireless technologies. Also directs NSF to develop a plan that describes the current status of broadband access for scientific research purposes. Introduced May 10, 2007; referred to House Committee on Science and Technology. Passed House May 21, 2007. Passed Senate July 19, 2007. Signed into law, August 9, 2007.

P.L. 110-161 (H.R. 2764)

Consolidated Appropriations Act, 2008. For Rural Utilities Service, U.S. Department of Agriculture, provides $6.45 million to support a loan level of $300 million for the broadband loan program, and $13.5 million for broadband community connect grants. For the USF extends for one year (until December 31, 2008) the USF exemption for the Antideficiency Act (Title V, Sec. 510); prohibits the FCC from using its FY2008 funds to limit USF support to a primary, or single, line (Title V, Sec. 511); permits the transfer of up to $21,480,000 of FY2008 funds from the USF to monitor the Program to prevent and remedy fraud, waste, and abuse, and to conduct audits and investigations by the OIG (Title V, FCC Salaries and Expenses). Signed by President, December 26, 2007.

P.L. 110-234 (H.R. 2419)

Food, Conservation, and Energy Act of 2008. Reauthorizes broadband program at the Rural Utilities Service through FY2012. Establishes a National Center for Rural Telecommunications Assessment. Directs USDA and the FCC to submit to Congress a comprehensive rural broadband strategy. Introduced May 22, 2007; referred to Committee on Agriculture, and in addition to Committee on Foreign Affairs. Subcommittee on Specialty Crops, Rural Development, and Foreign Agriculture held markup of Title VII (Rural Development) on June 6, 2007. Reported by House Committee on Agriculture (H.Rept. 110-256) on July 23, 2007. Passed House July 27, 2007. Passed Senate with an amendment, December 14, 2007. Conference report (H.Rept. 110-627) approved by the House May 14, 2008, and by the Senate May 15, 2008. Vetoed by the President, May 21, 2008. House and Senate overrode veto on May 21 and May 22, 2008. Became P.L. 110-234.

H.R. 42 (Velazquez)

Serving Everyone with Reliable, Vital Internet, Communications and Education Act of 2007. Directs the FCC to expand assistance provided by the Lifeline Assistance Program and the Link Up Program to include broadband service. Introduced January 4, 2007; referred to Committee on Energy and Commerce.

H.R. 278 (Cubin)

Amends Section 254 of the Communications Act of 1934 to provide that funds received as universal service contributions and the universal service support programs established pursuant to that section are not subject to certain provisions of title 31, United States Code, commonly known as the Antideficiency Act. Introduced January 5, 2007; referred to Committee on Energy and Commerce.

H.R. 1818 (Matsui)

Broadband Deployment Acceleration Act of 2007. Amends the Internal Revenue Code of 1986 to provide for the expensing of broadband Internet access expenditures. Introduced March 29, 2007; referred to Committee on Ways and Means.

H.R. 2035 (Herseth Sandlin)

Rural Broadband Improvement Act. Amends the Rural Electrification Act of 1936 to modify the broadband loan program at the Rural Utilities Service by narrowing the definition of "eligible rural community" and by limiting loans awarded to applicants proposing to serve areas that already have a broadband provider. Introduced April 25, 2007; referred to Committee on Agriculture and to Committee on Energy and Commerce.

H.R. 2054 (Boucher)

Universal Reform Act of 2007. Targets universal service support specifically to eligible telecommunications carriers in high-cost geographic areas to ensure that communications services and high-speed broadband services are made available throughout all of the States of the United States in a fair and equitable manner. Introduced April 26, 2007; referred to Committee on Energy and Commerce.

H.R. 2174 (Salazar)

Rural Broadband Initiative Act of 2007. Establishes an Office of Rural Broadband Initiatives within the Department of Agriculture which will administer all rural broadband grant and loan programs previously administered by the Rural Utilities Service. Also establishes a National Rural Broadband Innovation Fund which would fund experimental and pilot rural broadband projects and applications. Introduced May 3, 2007; referred to Committee on Agriculture and to Committee on Energy and Commerce.

H.R. 2569 (Graves)

Rural Broadband Deployment Act. Codifies certain changes proposed by USDA to the rules governing eligibility for the rural broadband access program. Specifically, would relax market survey requirements and eliminate the credit support requirement, including the cash-on-hand requirement. Introduced June 5, 2007; referred to Committee on Agriculture, and in addition to the Committee on Energy and Commerce.

H.R. 2829 (Serrano)

Financial Services and General Government Appropriations Act, 2008. The Senate Appropriations Committee-passed version of this 2008 appropriations bill includes language in Title V (sec. 501) to extend the FCC's universal service fund exemption for the Anti-deficiency Act until December 31, 2008, and includes language (sec. 502) to prohibit the FCC from implementing a single line restriction for universal service support. Passed by the Senate Appropriations Committee July 12, 2007, and reported out of Committee (S.Rept. 110-129) on July 13, 2007.

H.R. 2953 (Space)

Rural Broadband Access Enhancement Act. Seeks to redefine "eligible rural community," streamline application process and lower equity requirements, restrict loans to communities with existing broadband providers, eliminate limitation on eligibility based on number of subscriber lines, set 35-year maximum on term of loan repayment, and direct USDA/RUS to meet specific reporting requirements. Introduced July 10, 2007; referred to Committee on Agriculture and Committee on Energy and Commerce.

H.R. 3281 (Boucher)

Community Broadband Act of 2007. Sets forth that no state regulation or requirement shall prevent a public provider from offering broadband services, and prohibits a municipality from discriminating against competing private providers. Introduced August 1, 2007; referred to Committee on Energy and Commerce.

H.R. 3246 (Oberstar)

Regional Economic and Infrastructure Development Act of 2007. Designates five regional commissions throughout the U.S. which would provide economic and infrastructure development grants, including grants to develop the telecommunications infrastructure of the region. Introduced July 31, 2007; referred to Committee on Transportation and Infrastructure and to Committee on Financial Services. Reported by Committee on Transportation and Infrastructure, September 7, 2007 (H.Rept. 110-321, Part I). Passed by House, October 4, 2007.

H.R. 3428 (McHugh)

Rural America Digital Accessibility Act. Provides for grants, loan guarantees, research, and tax credits to promote broadband deployment in underserved rural areas. Introduced

August 3, 2007; referred to Committee on Energy and Commerce and in addition to the Committee on Ways and Means and the Committee on Science and Technology.

H.R. 3627 (Space)

Connect the Nation Act. Establishes a State Broadband Data and Development Grant Program within the Department of Commerce to help states develop and implement statewide initiatives to identify and track the availability and adoption of broadband services within each state. Authorizes $40 million for each of fiscal years 2008 through 2012. Introduced September 20, 2007; referred to Committee on Energy and Commerce.

H.R. 3893 (Allen)

Connect America Now Act. Establishes a State Broadband Data and Development Grant Program within the Department of Commerce to help states develop and implement statewide initiatives to identify and track the availability and adoption of broadband services within each state. Authorizes $40 million for each of fiscal years 2008 through 2012. Introduced October 18, 2007; referred to Committee on Energy and Commerce.

H.R. 3919 (Markey)

Broadband Census of America Act of 2007. Provides for a comprehensive inventory of existing broadband service. Directs the FCC to conduct an annual assessment of broadband deployment, including information on bandwidth service tiers, types of technology, and international comparisons. Directs NTIA to develop and maintain a broadband inventory map of the United States that depicts broadband deployment at a nine digit zip code area level, census tract level, or functional equivalent. Directs NTIA to award grants to states for broadband map development and grants for demand-side broadband service identification and assessments. Directs the FCC to conduct periodic consumer surveys of broadband service capability. Authorizes $20 million for each of fiscal years 2008 through 2010, of which not less than $15 million would be available for the state broadband map grants. Authorizes $50 million in FY2008, $100 million in FY2009, and $125 million in FY2010 for the demand-side broadband service identification and assessment (local technology planning) grants. Introduced October 22, 2007; referred to Committee on Energy and Commerce. Reported by Committee on Energy and Commerce (H.Rept. 110-443), November 13, 2007. Passed House by voice vote, November 13, 2007.

H.R. 5353 (Markey)

Internet Freedom Preservation Act of 2008. To establish broadband policy and direct the Federal Communications Commission to conduct a proceeding and public broadband summits to assess competition, consumer protection, and consumer choice issues relating to broadband Internet access, and for other purposes. Introduced February 12, 2008; referred to Committee on Energy and Commerce.

H.R. 5682 (Allen)

Rural America Communication Expansion for the Future Act of 2008. Reforms and reauthorizes through FY20 13 the Rural Broadband Access Loan and Loan Guarantee Program and the Community Connect Grant Program. Provides for tax incentives and NTIA grant program for broadband services in rural and underserved areas. Introduced April 2, 2008; referred to Committee on Energy and Commerce and in addition to the Committees on Ways and Means and Agriculture.

H.R. 5994 (Conyers)

Internet Freedom and Nondiscrimination Act of 2008. To amend the Clayton Act with respect to competitive and nondiscriminatory access to the Internet. Introduced May 8, 2008; referred to Committee on the Judiciary.

H.R. 6320 (Markey)

Twenty-first Century Communications and Video Accessability Act of 2008. Ensures that individuals with disabilities have access to emerging Internet Protocol- based communication and video programming technologies in the 21st Century. Introduced June 19, 2008; referred to Committee on Energy and Commerce.

H.R. 6356 (Barton)

The Universal Service Reform, Accountability, and Efficiency Act of 2008. Reforms the collection and distribution of universal service support under the Communications Act of 1934. Introduced June 24, 2008; referred to Committee on Energy and Commerce.

H.Res. 1292 (Eshoo)

Establishes a national goal for the universal deployment of next-generation broadband networks by 2015 and calls upon the Congress and the President to develop a strategy, enact legislation, and adopt policies to accomplish this objective. Introduced June 20, 2008; referred to Committee on Energy and Commerce.

S. 101 (Stevens)

Universal Service for Americans Act ("USA Act"). Directs the FCC to establish Broadband for Unserved Area Areas Program to be funded by the Universal Service Fund. Requires communications carriers to submit detailed broadband deployment data to the FCC. Introduced January 4, 2007; referred to Committee on Commerce, Science, and Transportation.

S. 215 (Dorgan)

Amend the Communications Act of 1934 to ensure net neutrality. Introduced January 9, 2007; referred to Committee on Commerce, Science, and Transportation.

S. 541 (Feingold)

Rural Opportunities Act of 2007. Directs the FCC to collect more detailed broadband deployment data and to periodically revise its definition of broadband above 200 kbps. Directs the Secretary of Agriculture to report on the adoption or planned adoption of the recommendations contained in the September 2005 audit report by the Inspector General of the United States Department of Agriculture. Introduced February 8, 2007; referred to Committee on Agriculture, Nutrition and Forestry.

S. 609 (Rockefeller)

A bill to amend Section 254 of the Communications Act of 1934 to provide that funds received as universal service contributions and the universal service support programs established pursuant to that section are not subject to certain provisions of Title 31, United States Code, commonly known as the Antideficiency Act. Introduced February 15, 2007; referred to the Committee on Commerce, Science, and Transportation.

S. 711 (Smith)

Universal Service for the 21st Century Act. Expands the contribution base for universal service and establishes a separate account within the universal service fund to support the deployment of broadband service in unserved areas. Introduced February 28, 2007; referred to Committee on Commerce, Science, and Transportation.

S. 761 (Reid)

America COMPETES Act. Authorizes the National Science Foundation (NSF) to provide grants for basic research in advanced information and communications technologies. Areas of research include affordable broadband access, including wireless technologies. Also directs NSF to develop a plan that describes the current status of broadband access for scientific research purposes. Introduced March 5, 2007; placed on Senate Legislative Calendar. Passed Senate April 25, 2007. Senate incorporated this measure in H.R. 2272 as an amendment July 19, 2007.

S. 1032 (Clinton)

Rural Broadband Initiative Act of 2007. Establishes an Office of Rural Broadband Initiatives within the Department of Agriculture which will administer all rural broadband grant and loan programs previously administered by the Rural Utilities Service. Also establishes a National Rural Broadband Innovation Fund which would fund experimental and pilot rural broadband projects and applications.

Introduced March 29, 2007; referred to Committee on Agriculture, Nutrition, and Forestry.

S. 1190 (Durbin)

Connect the Nation Act. Establishes a State Broadband Data and Development Grant Program within the Department of Commerce to help states develop and implement statewide initiatives to identify and track the availability and adoption of broadband services within each state. Authorizes $40 million for each of fiscal years 2008 through 2012. Introduced April 24, 2007; referred to Committee on Commerce, Science, and Transportation.

S.Res. 191 (Rockefeller)

Establishing a national goal for the universal deployment of next-generation broadband networks by 2015, and calling upon Congress and the President to develop a strategy, enact legislation, and adopt policies to accomplish this objective. Introduced May 8, 2007; referred to Committee on Commerce, Science, and Transportation.

S. 1264 (Coleman)

Rural Renaissance Act. Creates a Rural Renaissance Corporation which would fund qualified projects including projects to expand broadband technology in rural areas. Introduced May 2, 2007; referred to Committee on Finance.

S. 1439 (Roberts)

Rural Broadband Improvement Act of 2007. Reauthorizes the broadband and broadband loan guarantee program under Title VI of the Rural Electrification Acct of 1936. Introduced May 21, 2007; referred to Committee on Agriculture, Nutrition, and Forestry.

S. 1492 (Inouye)

Broadband Data Improvement Act. Seeks to improve the quality of federal broadband data collection and encourage state initiatives that promote broadband deployment. Directs the FCC to reevaluate its current 200 kbps broadband standard and to develop a new metric for “second generation broadband” capable of transmitting high definition video content. Directs broadband providers to report to the FCC connections within nine digit (zip+4) zip code areas. Directs the FCC to conduct its Section 706 inquiry into the status of broadband deployment on an annual basis. Directs the Census Bureau to collect residential broadband data. Directs GAO to develop broadband metrics involving connection cost and capability information that could be used to improve the process of comparing U.S. broadband deployment with other countries. Directs the Small Business Administration to conduct a study evaluating the impact of broadband speed and cost on small businesses. Authorizes $40 million for each of fiscal years 2008 through 2012 to establish a State Broadband Data and Development Grant Program within the Department of Commerce to help states develop and implement statewide initiatives to identify and track the availability and adoption of broadband services within each state. Introduced May 24, 2007; referred to Committee on Commerce, Science, and Transportation. Ordered to be reported July 19, 2007. Ordered to be reported July 19, 2007; reported by Committee (S.Rept. 110-204) and placed on Senate Legislative Calendar, October 24, 2007.

S. 1853 (Lautenberg)

Community Broadband Act of 2007. Sets forth that no state regulation or requirement shall prevent a public provider from offering broadband services, and prohibits a municipality from discriminating against competing private providers. Introduced July 23, 2007; referred to Committee on Commerce, Science, and Transportation. Ordered to be reported favorably with amendments by the Committee on Commerce, Science and Transportation, October 30, 2007.

S. 2242 (Baucus)

Heartland, Habitat, Harvest, and Horticulture Act of 2007. Introduced October 25, 2007; referred to Committee on Finance. Amends the Internal Revenue Code of 1986 to provide for the expensing of broadband Internet access expenditures. Creates a Rural Renaissance Corporation which would fund qualified projects including projects to expand broadband technology in rural areas. Reported to Senate (S.Rept. 110-206) and placed on Senate Legislative Calendar, October 25, 2007.

S. 2302 (Harkin)

Food and Energy Security Act of 2007. Reauthorizes broadband program at the Rural Utilities Service through FY20 12. Introduced November 2, 2007. Senate Committee on Agriculture, Nutrition, and Forestry reported measure to Senate (S.Rept. 110-220) November 2, 2007; placed on Senate Legislative Calendar.

S. 3182 (Mikulski)

Departments of Commerce and Justice, Science, and Related Agencies Appropriations Act, 2009. Provides $20 million to the Technology Opportunities Program (TOP) at the National Telecommunications and Information Administration, Department of Commerce, for competitive grants for the construction of broadband services and networks. Reported by Committee on Appropriations June 23, 2008 (S.Rept. 110-397).

S. 3260 (Durbin)

Financial Services and General Government Appropriations for FY2009. Makes $3 million available to the FCC to establish and administer a State Broadband Data and Development matching grants program for State-level broadband demand aggregation activities and creation of geographic inventory maps of broadband service to identify gaps in service and provide a baseline assessment of statewide broadband deployment. Reported by Committee on Appropriations July 14, 2008 (S.Rept. 110-417).

S. 3289 (Kohl)

Agriculture, Rural Development, Food and Drug Administration, and Related Agencies Appropriations, 2009. Provides $1 1.618 million to support a loan level of $297.9 million for the broadband loan program, and $ 13.406 million for broadband grants. Introduced July 21, 2008; referred to Committee on Appropriations. Reported to Senate (S.Rept. 110-426) July 21, 2008; placed on Senate Legislative Calendar.

S. 3297 (Reid)

Advancing America's Priorities Act. Title V, Subtitle A, Part 1 is the Broadband Data Improvement Act, which seeks to improve the quality of federal broadband data collection and encourage state initiatives that promote broadband deployment. Introduced July 22, 2008; placed on Senate Legislative Calendar.

ENDNOTES

[1] For further information, see CRS Report RS20993, *Wireless Technology and Spectrum Demand: Advanced Wireless Services,* by Linda K. Moore.

[2] For further information, see CRS Report RL32421, *Broadband Over Power Lines: Regulatory and Policy Issues,* by Patricia Moloney Figliola.

[3] FCC, *High-Speed Services for Internet Access: Status as of June 30, 2007,* March 2008. Available at [http://hraunfoss.fcc.gov/edocs_public/attachmatch/DOC-280906A1 .pdf]

[4] Percentage assumes one high speed line per household, 65.9 million residential high speed lines (per June 30, 2007 FCC data) and 114 million households in the U.S. (2006 Census data, see [http://www.census.gov/prod/2007pubs/08abstract/pop.pdf]).

[5] Federal Communications Commission, *Fourth Report to Congress*, "Availability of Advanced Telecommunications Capability in the United States," GN Docket No. 04-54, FCC 04-208, September 9, 2004, p. 38. Available at [http://hraunfoss.fcc.gov/edocs _public/attachmatch/FCC-04-208A1 .pdf]

[6] International Telecommunications Union, *Economies by broadband penetration, 2007.* Available at [http://www.itu.int/ITU-D/ict/statistics/at_glance/top20_broad_2007.html].

[7] OECD, *OECD Broadband Statistics, December 2007.* Available at [http:// www.oecd.org/sti/ict/broadband].

[8] OECD, Directorate for Science, Technology and Industry, *The Development of Broadband Access in OECD Countries*, October 29, 2001, 63 pages. For a comparison of government broadband policies, also see OECD, Directorate for Science, Technology and Industry, *Broadband Infrastructure Deployment: The Role of Government Assistance*, May 22, 2002, 42 p.

[9] See Turner, Derek S., Free Press, *Broadband Reality Check II: The Truth Behind America's Digital Divide*, August 2006, pp 8-11. Available at [http://www.freepress.net/ files/bbrc2-final.pdf]; and Turner, Derek S., Free Press, *'Shooting the Messenger' Myth vs. Reality: U.S. Broadband Policy and International Broadband Rankings*, July 2007, 25 p., available at [http://www.freepress.net/files/shooting_the_messenger.pdf].

[10] National Telecommunications and Information Administration, *Fact Sheet: United States Maintains Information and Communication Technology (ICT) Leadership and Economic Strength*, available at [http: //www.ntia.doc .gov/ntiahome/press/2007/ICTleader _042407.html].

[11] See Wallsten, Scott, Progress and Freedom Foundation, *Towards Effective U.S. Broadband Policies*, May 2007, 19 pages. Available at [http://www.pff.org/issues-pubs/pops/pop 14.7usbroadbandpolicy.pdf]. Also see Ford, George, Phoenix Center, *The Broadband Performance Index: A Policy-Relevant Method of Comparing Broadband Adoption Among Countries*, Phoenix Center Policy Paper Number 29, July 2007, 32 pp. Available at [http://www.phoenix-center.org/pcpp/PCPP29Final.pdf].

[12] See price and services and speed data on OECD Broadband Portal, available at [http://www.oecd.org/sti/ict/broadband]; Turner, Derek S., Free Press, *Broadband Reality*

Check II: The Truth Behind America's Digital Divide, August 2006, pp 5-9; Kende, Michael, Analysis Consulting Limited, *Survey of International Broadband Offerings*, October 4, 2006, 12 pages, available at [http://www.analysys.com/pdfs/Broadband PerformanceSurvey.pdf]; and Correa, Daniel K., The International Technology and Innovation Foundation, *Assessing Broadband in America: OECD and ITIF Broadband Rankings*, April 2007, 10 pages, available at [http://www.itif.org/files/ Broadband Rankings.pdf].

[13] For more information on broadband and the digital divide, see CRS Report RL307 19, *Broadband Internet Access and the Digital Divide: Federal Assistance Programs,* by Lennard G. Kruger and Angele A. Gilroy.

[14] FCC, *High-Speed Services for Internet Access: Status as of June 30, 2007,* p.4.

[15] Horrigan, John B., Pew Internet & American Life Project, *Home Broadband Adoption 2008,* July 2008, p. 3. Available at [http://www.pewinternet.org/pdfs/PIP_Broadband_ 2008.pdf]

[16] *Fourth Report*, p. 8.

[17] See Appendix C of the *Fourth Report*, "List of Broadband-Related Proceedings at the Commission," pp. 54-56.

[18] Ibid., p. 9.

[19] Ibid., p. 5, 7.

[20] U.S. Government Accountability Office, *Broadband Deployment is Extensive throughout the United States, but It Is Difficult to Assess the Extent of Deployment Gaps in Rural Areas*, GAO-06-426, May 2006, p. 3.

[21] FCC, News Release, "FCC Expands, Improves Broadband Data Collection," March 19, 2008. Available at [http://hraunfoss.fcc.gov/edocs_public/attachmatch/DOC-280909A1 .pdf].

[22] See speech by Nancy Victory, Assistant Secretary for Communications and Information, before the National Summit on Broadband Deployment, October 25, 2001, [http://www.ntia.doc.gov/ntiahome/speeches/200 1/broadband_1 02501 .htm].

[23] Address by Nancy Victory, NTIA Administrator, before the Alliance for Public Technology Broadband Symposium, February 8, 2002, [http://www.ntia.doc.gov/ ntiahome/speeches/2002/apt_020802.htm].

[24] Allen, Mike, "Bush Sets Internet Access Goal," *Washington Post*, March 27, 2004.

[25] See White House, *A New Generation of American Innovation*, April 2004. Available at [http://www.whitehouse.gov/infocus/technology/economic_policy200404/innovation. pdf].

[26] Available at [http://www.ntia.doc.gov/reports/2008/NetworkedNationBroadbandin America2007.pdf]

[27] NTIA, *Press Release*, "Gutierrez Hails Dramatic U.S. Broadband Growth," January 31, 2008. Available at [http://www.ntia.doc.gov/ntiahome/press/2008/NetworkedNation_ 013108.html].

[28] For more information on the RUS broadband programs, see CRS Report RL33816, *Broadband Loan and Grant Programs in the USDA's Rural Utilities Service,* by Lennard G. Kruger.

[29] For further information see CRS Report RL32949, *Communications Act Revisions: Selected Issues for Consideration*, Angele A. Gilroy, coordinator.

[30] For further information on the net neutrality debate, see CRS Report RS22444, *Net Neutrality: Background and Issues*, by Angele A. Gilroy.

[31] For further information on the Universal Service Fund and related FCC and congressional activity see CRS Report RL33979, *Universal Service Fund: Background and Options for Reform*, by Angele A. Gilroy.

[32] For further details, see CRS Report RL33816, *Broadband Loan and Grant Programs in the USDA's Rural Utilities Service, by* Lennard G. Kruger.

In: Maximizing Broadband Services to Rural Communities ISBN 978-1-60692-846-2
Editors: V.B. Haynesworth, A.R. Harrison, pp.35-67

Chapter 3

BROADBAND INTERNET ACCESS AND THE DIGITAL DIVIDE: FEDERAL ASSISTANCE PROGRAMS*

Lennard G. Kruger and Angele A. Gilroy

ABSTRACT

The "digital divide" is a term that has been used to characterize a gap between "information haves and have-nots," or in other words, between those Americans who use or have access to telecommunications technologies (e.g., telephones, computers, the Internet) and those who do not. One important subset of the digital divide debate concerns high-speed Internet access, also known as *broadband*. Broadband is provided by a series of technologies (e.g., cable, telephone wire, fiber, satellite, wireless) that give users the ability to send and receive data at volumes and speeds far greater than current "dial-up" Internet access over traditional telephone lines.

Broadband technologies are currently being deployed primarily by the private sector throughout the United States. While the numbers of new broadband subscribers continue to grow, studies and data suggest that the rate of broadband deployment in urban and high income areas may be outpacing deployment in rural and low-income areas. Some policymakers, believing that disparities in broadband access across American society could have adverse economic and social consequences on those left behind, assert that the federal government should play a more active role to avoid a "digital divide" in broadband access. One approach is for the federal government to provide financial assistance to support broadband deployment in underserved areas. Others, however, believe that federal assistance for broadband deployment is not appropriate. Some opponents question the reality of the "digital divide," and argue that federal intervention in the broadband marketplace would be premature and, in some cases, counterproductive.

Legislation introduced in the 1 10th Congress seeks to provide federal financial assistance for broadband deployment in the form of grants, loans, subsidies, and tax credits. Of particular note is the reauthorization and reform of the Rural Utilities Service broadband loan program, which was enacted as part of the 2008 farm bill (P.L. 110-234).

* Excerpted from CRS Report for Congress, Order Code RL30719, dated August 22, 2008.

Legislation to reform universal service — which could impact the amount of financial assistance available for broadband deployment in rural areas — has been introduced (H.R. 42, H.R. 2054, H.R. 3428, H.R. 6320, H.R. 6356, S. 101, S. 711), as well as other broadband legislation (H.R. 1818, H.R. 2035, H.R. 2174, H.R. 2272, H.R. 2419, H.R. 2569, H.R. 2720, H.R. 2764, H.R. 2953, H.R. 3246, H.R. 3281, H.R. 3627, H.R. 3893, H.R. 3919, H.R. 5682, S. 541, S. 761, S. 1032, S. 1190, S. 1264, S. 1439, S. 1492, S. 2242, S. 2302).

In assessing such legislation, several policy issues arise. For example, is the current status of broadband deployment data an adequate basis on which to base policy decisions? Is federal assistance premature, or do the risks of delaying assistance to underserved areas outweigh the benefits of avoiding federal intervention in the marketplace? And finally, if one assumes that governmental action is necessary to spur broadband deployment in underserved areas, which specific approaches, either separately or in combination, are likely to be most effective?

BACKGROUND

The "digital divide" is a term used to describe a perceived gap between perceived "information haves and have-nots," or in other words, between those Americans who use or have access to telecommunications technologies (e.g., telephones, computers, the Internet) and those who do not.[1] Whether or not individuals or communities fall into the "information haves" category depends on a number of factors, ranging from the presence of computers in the home, to training and education, to the availability of affordable Internet access. A series of reports issued by the Department of Commerce[2] (DOC) during the Clinton Administration argued that a "digital divide" exists, with many rural citizens, certain minority groups, and low-income Americans tending to have less access to telecommunications technology than other Americans.[3]

In February 2002, the Bush Administration's Department of Commerce released its first survey report on Internet use, entitled *A Nation Online: How Americans Are Expanding Their Use of the Internet.*[4] While acknowledging a disparity in usage between "information haves and have nots," the report focused on the increasing rates of Internet usage among traditionally underserved groups:

> In every income bracket, at every level of education, in every age group, for people of every race and among people of Hispanic origin, among both men and women, many more people use computers and the Internet now than did so in the recent past. Some people are still more likely to be Internet users than others. Individuals living in low-income households or having little education, still trail the national average. However, broad measures of Internet use in the United States suggest that over time Internet use has become more equitable.[5]

A Nation Online: Entering the Broadband Age, published in September 2004, was the sixth Department of Commerce report examining the use of computers, the Internet, and other information technology. For the first time, the DOC report focused on broadband, also known as high-speed Internet access. Broadband is provided by a series of technologies (e.g., cable, telephone wire, satellite, wireless) that give users the ability to send and receive data at volumes and speeds far greater than current "dial-up" Internet access over traditional telephone lines.[6]

According to the latest FCC data on the deployment of high-speed Internet connections (released March 2008), as of June 30, 2007, there were 100.9 million high speed lines connecting homes and businesses to the Internet in the United States, a growth rate of 22% during the first half of 2007. Of the 100.9 million high speed lines reported by the FCC, 65.9 million serve residential users.[7] While the broadband *adoption* rate stands at roughly 58% of U.S. households,[8] broadband *availability* is much higher. As of June 30, 2007, the FCC found at least one high-speed subscriber in 99% of all zip codes in the United States. The FCC has estimated that "roughly 20 percent of consumers with access to advanced telecommunications capability do subscribe to such services." According to the FCC, possible reasons for the gap between broadband availability and subscribership include the lack of computers in some homes, price of broadband service, lack of content, and the availability of broadband at work.[9]

Broadband in Rural and Underserved Areas[10]

While the number of new broadband subscribers continues to grow, the rate of broadband deployment in urban and high income areas appears to be outpacing deployment in rural and low-income areas. In response to a request by ten Senators, the Departments of Commerce and Agriculture released a report on April 26, 2000, concluding that rural areas lag behind urban areas in access to broadband technology. The report found that less than 5% of towns of 10,000 or less have access to broadband, while broadband over cable has been deployed in more than 65% of all cities with populations over 250,000, and broadband over the telephone network has been deployed in 56% of all cities with populations over 100,000.[11]

More recently, a May 2006 report released by the Government Accountability Office (GAO) found that 17% of rural households subscribe to broadband, as opposed to 28% of suburban and 29% of urban households.[12]

Similarly, 2008 data from the Pew Internet & American Life Project indicate that while broadband adoption is growing in urban, suburban, and rural areas, broadband users make up larger percentages of urban and suburban users than rural users. Pew found that the percentage of all U.S. adults with broadband at home is 60% for suburban areas, 57% for urban areas, and 38% for rural areas.[13]

According to the latest FCC data on the deployment of high-speed Internet connections (released March 2008), high-speed subscribers were reported in 99% of the most densely populated zip codes, as opposed to 91% of zip codes with the lowest population densities. Similarly, for zip codes ranked by median family income, high-speed subscribers were reported present in 99% of the top one-tenth of zip codes, as compared to 92% of the bottom one-tenth of zip codes.[14]

The 2007 Department of Commerce report, *Networked Nation: Broadband in America 2007*,[15] found that a lower percentage of households have broadband in rural areas (3 8.8%) than in urban areas (53.7%). The report found that broadband penetration rates are higher in the West and Northeast than in the South and Midwest. Race and ethnicity were also found to be significant determinants of broadband use, with 54.9% of White households having broadband, as opposed to 3 6.4% of Black and 3 5.2% of Hispanic households.[16] Recent data from the Pew Internet and American Life Project also show gaps in broadband adoption by race, ethnicity, income, and education level.[17]

Some policymakers believe that disparities in broadband access across American society could have adverse consequences on those left behind. While a minority of American homes today subscribe to broadband, many believe that advanced Internet applications of the future — voice over the Internet protocol (VoIP) or high quality video, for example — and the resulting ability for businesses and consumers to engage in e-commerce, may increasingly depend on high speed broadband connections to the Internet. Thus, some say, communities and individuals without access to broadband could be at risk to the extent that e-commerce becomes a critical factor in determining future economic development and prosperity. A 2003 study conducted by Criterion Economics found that ubiquitous adoption of current generation broadband technologies would result in a cumulative increase in gross domestic product of $179.7 billion, while sustaining an additional 61,000 jobs per year over the next nineteen years. The study projected that 1.2 million jobs could be created if next generation broadband technology is rapidly and ubiquitously deployed.[18] A February 2006 study done by the Massachusetts Institute of Technology for the Economic Development Administration/ Department of Commerce marked the first attempt to quantitatively measure the impact of broadband on economic growth. The study found that "between 1998 and 2002, communities in which mass-market broadband was available by December 1999 experienced more rapid growth in employment, the number of businesses overall, and businesses in IT-intensive sectors, relative to comparable communities without broadband at that time."[19]

Subsequently, a June 2007 report from the Brookings Institution found that for every one percentage point increase in broadband penetration in a state, employment is projected to increase by 0.2 to 0.3% per year. For the entire U.S. private non-farm economy, the study projected an increase of about 300,000 jobs, assuming the economy is not already at full employment.[20]

Some also argue that broadband is an important contributor to U.S. future economic strength with respect to the rest of the world. According to the International Telecommunications Union, the U.S. ranks 24th worldwide in broadband penetration (subscriptions per 100 inhabitants in 2007).[21] Data from the Organization for Economic Cooperation and Development (OECD) found the U.S. ranking 15th among OECD nations in broadband access per 100 inhabitants as of December 2007.[22] By contrast, in 2001 an OECD study found the U.S. ranking 4th in broadband subscribership per 100 inhabitants (after Korea, Sweden, and Canada).[23] While many argue that the U.S. declining performance in international broadband rankings is a cause for concern,[24] others — including the Administration — maintain that the OECD and ITU data undercount U.S. broadband deployment,[25] and that cross-country broadband deployment comparisons are not necessarily meaningful and inherently problematic.[26] Finally, an issue related to international broadband rankings is the extent to which broadband speeds and prices differ between the U.S. and the rest of the world.[27]

Federal Role

The Telecommunications Act of 1996 (P.L. 104-104) addresses the issue of whether the federal government should intervene to prevent a "digital divide" in broadband access.

Section 706 requires the FCC to determine whether "advanced telecommunications capability [i.e., broadband or high-speed access] is being deployed to all Americans in a reasonable and timely fashion." If this is not the case, the act directs the FCC to "take immediate action to accelerate deployment of such capability by removing barriers to infrastructure investment and by promoting competition in the telecommunications market."

On January 28, 1999, the FCC adopted its first report (FCC 99-5) pursuant to Section 706. The report concluded that "the consumer broadband market is in the early stages of development, and that, while it is too early to reach definitive conclusions, aggregate data suggests that broadband is being deployed in a reasonable and timely fashion."[28] The FCC announced that it would continue to monitor closely the deployment of broadband capability in annual reports and that, where necessary, it would "not hesitate to reduce barriers to competition and infrastructure investment to ensure that market conditions are conducive to investment, innovation, and meeting the needs of all consumers."

The FCC's second Section 706 report was adopted on August 3, 2000. Based on more extensive data than the first report, the FCC similarly concluded that notwithstanding risks faced by some vulnerable populations, broadband is being deployed in a reasonable and timely fashion overall:

> Recognizing that the development of advanced services infrastructure remains in its early stages, we conclude that, overall, deployment of advanced telecommunications capability is proceeding in a reasonable and timely fashion. Specifically, competition is emerging, rapid build-out of necessary infrastructure continues, and extensive investment is pouring into this segment of the economy.[29]

The FCC's third Section 706 report was adopted on February 6, 2002. Again, the FCC concluded that "the deployment of advanced telecommunications capability to all Americans is reasonable and timely."[30] The FCC added:

> We are encouraged by the expansion of advanced services to many regions of the nation, and growing number of subscribers. We also conclude that investment in infrastructure for most advanced services markets remains strong, even though the pace of investment trends has generally slowed. This may be due in part to the general economic slowdown in the nation. In addition, we find that emerging technologies continue to stimulate competition and create new alternatives and choices for consumers.[31]

On September 9, 2004, the FCC adopted and released its *Fourth Report* pursuant to Section 706. Like the previous three reports, the FCC concluded that "the overall goal of Section 706 is being met, and that advanced telecommunications capability is indeed being deployed on a reasonable and timely basis to all Americans."[32] The FCC noted the emergence of new services such as VoIP, and the significant development of new broadband access technologies such as unlicensed wireless (WiFi) and broadband over power lines. The FCC noted the future promise of emerging multiple advanced broadband networks which can complement one another:

> For example, in urban and suburban areas, wireless broadband services may "fill in the gaps" in wireline broadband coverage, while wireless and satellite services may bring high-speed broadband to remote areas where wireline deployment may be costly. Having multiple

advanced networks will also promote competition in price, features, and quality-of-service among broadband-access providers.[33]

Two FCC Commissioners (Michael Copps and Jonathan Adelstein) dissented from the *Fourth Report* conclusion that broadband deployment is reasonable and timely. They argued that the relatively poor world ranking of United States broadband penetration indicates that deployment is insufficient, that the FCC's continuing definition of broadband as 200 kilobits per second is outdated and is not comparable to the much higher speeds available to consumers in other countries, and that the use of zip code data (measuring the presence of at least one broadband subscriber within a zip code area) does not sufficiently characterize the availability of broadband across geographic areas.[34]

On March 19, 2008, the FCC adopted the *Fifth Report* to Congress on broadband deployment under Section 706 of the Telecommunications Act of 1996. The *Fifth Report* was released on June 12, 2008.[35] As did previous reports, the *Fifth Report* found that broadband services are being deployed to all Americans in a reasonable and timely fashion. Commissioners Copps and Adelstein again dissented, citing flawed data collection methodologies, lagging U.S. broadband penetration internationally, and the lack of a comprehensive U.S. broadband strategy.

Meanwhile, the National Telecommunications and Information Administration (NTIA) at the Department of Commerce (DOC) was tasked with developing the Bush Administration's broadband policy.[36] Statements from Administration officials indicated that much of the policy would focus on removing regulatory roadblocks to investment in broadband deployment.[37] On June 13, 2002, in a speech at the 21st Century High Tech Forum, President Bush declared that the nation must be aggressive about the expansion of broadband, and cited ongoing activities at the FCC as important in eliminating hurdles and barriers to get broadband implemented. President Bush made similar remarks citing the economic importance of broadband deployment at the August 13, 2002 economic forum in Waco, Texas.

Subsequently, a more formal Administration broadband policy was unveiled in March and April of 2004. On March 26, 2004, President Bush endorsed the goal of universal broadband access by 2007. Then on April 26, 2004, President Bush announced a broadband initiative which includes promoting legislation which would permanently prohibit all broadband taxes, making spectrum available for wireless broadband and creating technical standards for broadband over power lines, and simplifying rights-of-way processes on federal lands for broadband providers.[38]

The Bush Administration has also emphasized the importance of encouraging demand for broadband services. On September 23, 2002, the DOC's Office of Technology Policy released a report, *Understanding Broadband Demand: A Review of Critical Issues* ,[39] which argued that national governments can accelerate broadband demand by taking a number of steps, including protecting intellectual property, supporting business investment, developing e-government applications, promoting efficient radio spectrum management, and others. Similarly, the President's Council of Advisers on Science & Technology (PCAST) was tasked with studying "demand- side" broadband issues and suggesting policies to stimulate broadband deployment and economic recovery. The PCAST report, *Building Out Broadband*, released in December 2002, concluded that while government should not intervene in the telecommunications marketplace, it should apply existing policies and work with the private sector to promote broadband applications and usage. Specific initiatives include increasing e-government broadband applications (including homeland security); promoting telework, distance learning, and telemedicine; pursuing broadband-friendly spectrum policies, and ensuring access to public rights of way for broadband infrastructure.[40]

On January 31, 2008, NTIA released a report, entitled, *Networked Nation:Broadband in America, 2007.*[41] According to NTIA, the report showed "that the Administration's technology, regulatory, and fiscal policies have stimulated innovation and competition, and encouraged investment in the U.S. broadband market contributing to significantly increased accessibility of broadband services."[42]

Some policymakers in Congress have disagreed with the Administration's assessment and assert that the federal government should play a more active role to avoid a "digital divide" in broadband access, and that legislation is necessary to ensure fair competition and timely broadband deployment. Bills have been introduced into past Congresses (and have been introduced in the 1 10th Congress) seeking to provide federal financial assistance for broadband deployment in the form of grants, loans, subsidies, and/or tax credits.

State and Local Broadband Activities

In addition to federal support for broadband deployment, there are programs and activities ongoing at the state and local level. Surveys, assessments, and reports from the California Public Utilities Commission,[43] the AEI-Brookings Joint Center,[44] the National Conference of State Legislatures,[45] and the National Governors Association[46] have explored state and local broadband programs. A related issue is the emergence of municipal broadband networks (primarily wireless and fiber based) and the debate over whether such networks constitute unfair competition with the private sector (for more information on municipal broadband, see CRS Report RS20993, *Wireless Technology and Spectrum Demand: Advanced Wireless Services*, by Linda K. Moore).

FEDERAL PROGRAMS SUPPORTING BROADBAND

The Rural Broadband Access Loan and Loan Guarantee Program and the Community Connect Broadband Grants, both at the Rural Utilities Service of the U.S. Department of Agriculture, are currently the only federal programs *exclusively* dedicated to deploying broadband infrastructure. However, there exist other federal programs that provide financial assistance for various aspects of telecommunications development. The major vehicle for funding telecommunications development, particularly in rural and low-income areas, is the Universal Service Fund (USF).

While the USF's High Cost Program does not *explicitly* fund broadband infrastructure, subsidies are used, in many cases, to upgrade existing telephone networks so that they are capable of delivering high-speed services. Additionally, subsidies provided by USF's Schools and Libraries Program and Rural Health Care Program are used for a variety of telecommunications services, including broadband access.

Table 1 shows selected federal domestic assistance programs throughout the federal government that can be associated with telecommunications development. Many (if not most) of these programs can be related, if not necessarily to the deployment of broadband technologies in particular, then to telecommunications and the "digital divide" issue generally.

Table 2presents selected federal programs that have provided financial assistance for broadband. These programs are broken down into three categories: first, programs that fund access to telecommunications services in unserved or underserved areas; second, general economic development programs that have funded broadband-related projects; and third, applications-specific programs which will typically fund some aspect of broadband access as a means towards supporting a particular application, such as distance learning or telemedicine.

The Universal Service Concept and the FCC[47]

Since its creation in 1934 the Federal Communications Commission (FCC) has been tasked with "... mak[ing] available, so far as possible, to all the people of the United States, ... a rapid, efficient, Nation-wide, and world-wide wire and radio communications service with adequate facilities at reasonable charges... ."[48] This mandate led to the development of what has come to be known as the universal service concept.

The universal service concept, as originally designed, called for the establishment of policies to ensure that telecommunications services are available to all Americans, including those in rural, insular and high cost areas, by ensuring that rates remain affordable. Over the years this concept fostered the development of various FCC policies and programs to meet this goal. The FCC offers universal service support through a number of direct mechanisms that target both providers of and subscribers to telecommunications services.[49]

The development of the federal universal service high cost fund is an example of provider-targeted support. Under the high cost fund, eligible telecommunications carriers, usually those serving rural, insular and high cost areas, are able to obtain funds to help offset the higher than average costs of providing telephone service.[50] This mechanism has been particularly important to rural America where the lack of subscriber density leads to significant costs. FCC universal service policies have also been expanded to target individual users. Such federal programs include two income-based programs, Link Up and Lifeline, established in the mid-1980s to assist economically needy individuals. The Link Up program assists low-income subscribers pay the costs associated with the initiation of telephone service and the Lifeline program assists low-income subscribers pay the recurring monthly service charges. Funding to assist carriers providing service to individuals with speech and/or hearing disabilities is also provided through the Telecommunications Relay Service Fund. Effective January 1, 1 998, schools, libraries, and rural health care providers also qualified for universal service support.

Universal Service and the Telecommunications Act of 1996

Passage of the Telecommunications Act of 1996 (P.L. 104-104) codified the long-standing commitment by U.S. policymakers to ensure universal service in the provision of telecommunications services.

The Schools and Libraries, and Rural Health Care Programs

Congress, through the 1 996 Act, not only codified, but also expanded the concept of universal service to include, among other principles, that elementary and secondary schools and classrooms, libraries, and rural health care providers have access to telecommunications services for specific purposes at discounted rates. (See Sections 254(b)(6) and 254(h)of the 1996 Telecommunications Act, 47 U.S.C. 254.)

1. The Schools and Libraries Program. Under universal service provisions contained in the 1 996 Act, elementary and secondary schools and classrooms and libraries are designated as beneficiaries of universal service discounts. Universal service principles detailed in Section 254(b)(6) state that "Elementary and secondary schools and classrooms ... and libraries should have access to advanced telecommunications services...." The act further requires in Section 254(h)(1)(B) that services within the definition of universal service be provided to elementary and secondary schools and libraries for education purposes at discounts, that is at "rates less than the amounts charged for similar services to other parties."

   The FCC established the Schools and Libraries Division within the Universal Service Administrative Company (USAC) to administer the schools and libraries or "E (education)-rate" program to comply with these provisions. Under this program, eligible schools and libraries receive discounts ranging from 20 to 90 percent for telecommunications services depending on the poverty level of the school's (or school district's) population and its location in a high cost telecommunications area. Three categories of services are eligible for discounts: internal connections (e.g., wiring, routers and servers); Internet access; and telecommunications and dedicated services, with the third category receiving funding priority. According to data released by program administrators, $21.3 billion in funding has been committed over the first ten years of the program with funding released to all states, the District of Columbia and all territories. Funding commitments for funding Year 2008 (July 1, 2008 - June 30, 2009), the eleventh and current year of the program, totaled $1.1 billion as of August 19, 2008.[51]
2. The Rural Health Care Program. Section 254(h) of the 1996 Act requires that public and non-profit rural health care providers have access to telecommunications services necessary for the provision of health care services at rates comparable to those paid for similar services in urban areas. Subsection 254(h)(1) further specifies that "to the extent technically feasible and economically reasonable" health care providers should have access to advanced telecommunications and information services. The FCC established the Rural Health Care Division (RHCD) within the USAC to administer the universal support program to comply with these provisions. Under FCC established rules only public or nonprofit health care providers are eligible to receive funding. Eligible health care providers, with the exception of those requesting only access to the Internet, must also be located in a rural area. The funding ceiling, or cap, for this support was established at $400 million annually. The funding level for Year One of the program (January 1998 - June 30, 1999) was set at $100 million. Due to less than anticipated demand, the FCC established a $12 million funding level for the second year (July 1, 1999 to June 30, 2000) of the program but has since returned to a $400 million yearly cap. As of December 31, 2007, covering

the first 10 years of the program, a total of $221.2 million has been committed to 3,784 rural health care providers. The primary use of the funding is to provide reduced rates for telecommunications and information services necessary for the provision of health care.[52]

The Telecommunications Development Fund

Section 714 of the 1996 Act created the Telecommunications Development Fund (TDF). The TDF is a private, non-governmental, venture capital corporation currently overseen by a five- member board of directors and fund management. The TDF focuses on seed, early stage, and select later stage investments in communications and has $90 million under management in two funds. Fund I, with a portfolio of $25 million invested in five companies, is no longer making new investments. Fund II which contains $65 million remains active and currently has 12 companies in its investment portfolio Funding is largely derived from the interest earned from the upfront payments bidders submit to participate in FCC auctions. The TDF also provides entrepreneur education, training, management and technical assistance in underserved rural and urban communities through the TDF Foundation.[53]

Universal Service and Broadband

One of the policy debates surrounding universal service is whether access to advanced telecommunications services (i.e. broadband) should be incorporated into universal service objectives. The term universal service, when applied to telecommunications, refers to the ability to make available a basket of telecommunications services to the public, across the nation, at a reasonable price. As directed in the 1996 Telecommunications Act [Section 254(c)] a federal-state Joint Board was tasked with defining the services which should be included in the basket of services to be eligible for federal universal service support; in effect using and defining the term "universal service" for the first time. The Joint Board's recommendation, which was subsequently adopted by the FCC in May 1997, included the following in its universal service package: voice grade access to and some usage of the public switched network; single line service; dual tone signaling; access to directory assistance; emergency service such as 911; operator services; access and interexchange (long distance) service.

Some policy makers expressed concern that the FCC-adopted definition is too limited and does not take into consideration the importance and growing acceptance of advanced services such as broadband and Internet access. They point to a number of provisions contained in the Universal Service section of the 1996 Act to support their claim. Universal service principles contained in Section 254(b)(2) state that "Access to advanced telecommunications services should be provided to all regions of the Nation." The subsequent principle (b)(3) calls for consumers in all regions of the nation including "low-income" and those in "rural, insular, and high cost areas" to have access to telecommunications and information services including "advanced services" at a comparable level and a comparable rate charged for similar services in urban areas. Such provisions, they state, dictate that the FCC expand its universal service definition.

Others caution that a more modest approach is appropriate given the "universal mandate" associated with this definition and the uncertainty and costs associated with mandating nationwide deployment of such advanced services as a universal service policy goal. Furthermore they state the 1996 Act does take into consideration the changing nature of the telecommunications sector and allows for the universal service definition to be modified if future conditions warrant. Section 254(c)of the act states that "universal service is an evolving level of telecommunications services" and the FCC is tasked with "periodically" reevaluating this definition "taking into account advances in telecommunications and information technologies and services." Furthermore, the Joint Board is given specific authority to recommend "from time to time" to the FCC modification in the definition of the services to be included for federal universal service support. The Joint Board, on November 19, 2007, concluded such an inquiry and recommended that the FCC change the mix of services eligible for universal service support. The Joint Board recommended, among other things, that "the universal availability of broadband Internet services" be included in the nation's communications goals and hence be supported by federal universal service funds.[54] In response to the Joint Board recommendation, the FCC, on January 29, 2008, released three notices of proposed rulemaking dealing with specific aspects of universal service, including an examination of the scope of the definition. The FCC has one year from the date of the issued Joint Board recommendation (November 20, 2008) to act on, but is not required to adopt, these recommendations.

Rural Utilities Service

The Rural Electrification Administration (REA), subsequently renamed the Rural Utilities Service (RUS), was established by the Roosevelt Administration in 1935. Initially, it was established to provide credit assistance for the development of rural electric systems. In 1949, the mission of REA was expanded to include rural telephone providers. Congress further amended the Rural Electrification Act in 1971 to establish within REA a Rural Telephone Account and the Rural Telephone Bank (RTB). Rural Telephone Loans and Loan Guarantees provide long-term direct and guaranteed loans for telephone lines, facilities, or systems to furnish and improve telecommunications service in rural areas. The RTB — liquidated in FY2006 — was a public-private partnership intended to provide additional sources of capital that would supplement loans made directly by RUS. Another program, the Distance Learning and Telemedicine Program, specifically addresses health care and education needs of rural America.

RUS implements two programs specifically targeted at providing assistance for broadband deployment in rural areas: the Rural Broadband Access Loan and Loan Guarantee Program and Community Connect Broadband Grants. The 110th Congress reauthorized and reformed the Rural Broadband Access Loan and Loan Guarantee program as part of the 2008 farm bill (P.L. 110-234). For further information on rural broadband and the RUS broadband programs, see CRS Report RL33816, *Broadband Loan and Grant Programs in the USDA's Rural Utilities Service*, by Lennard G. Kruger.

LEGISLATION IN THE 109TH CONGRESS

In the 1 09th Congress, legislation was introduced to provide financial assistance to encourage broadband deployment (including loans, grants, and tax incentives), and to allocate additional spectrum for use by wireless broadband applications. Of particular note was enactment of the Deficit Reduction Act of 2005 (P.L. 109-17 1), which set a hard deadline for the digital television transition, thereby reclaiming analog television spectrum to be auctioned for commercial applications such as wireless broadband.

LEGISLATION IN THE 110TH CONGRESS

In the 1 10th Congress, legislation has been introduced that would provide financial assistance for broadband deployment. Of particular note is the reauthorization of the Rural Utilities Service (RUS) broadband loan program, which was enacted as part of the 2008 farm bill (P.L. 110-234). In addition to reauthorizing and reforming the RUS broadband loan program, P.L. 110-234 contains provisions establishing a National Center for Rural Telecommunications Assessment and requiring the FCC and RUS to formulate a comprehensive rural broadband strategy.

Legislation to reform universal service (H.R. 2054, H.R. 6320, H.R. 6356, S. 101, S. 711) — which could have a significant impact on the amount of financial assistance available for broadband deployment in rural and underserved areas — has also been introduced. Additionally, Congress is considering broadband data bills — S. 1492 (reported by the Senate Commerce, Science, and Transportation Committee) and H.R. 3919 (passed by the House).

Meanwhile, the America COMPETES Act (H.R. 2272) was enacted (P.L. 110- 69) and contains a provision authorizing the National Science Foundation (NSF) to provide grants for basic research in advanced information and communications technologies. Areas of research include affordable broadband access, including wireless technologies. P.L. 110-69 also directs NSF to develop a plan that describes the current status of broadband access for scientific research purposes.

The following is a complete listing of bills in the 1 10th Congress.

P.L. 110-69 (H.R. 2272)

America COMPETES Act. Authorizes the National Science Foundation (NSF) to provide grants for basic research in advanced information and communications technologies. Areas of research include affordable broadband access, including wireless technologies. Also directs NSF to develop a plan that describes the current status of broadband access for scientific research purposes. Introduced May 10, 2007; referred to House Committee on Science and Technology. Passed House May 21, 2007. Passed Senate July 19, 2007. Signed into law, August 9, 2007.

P.L. 110-161 (H.R. 2764)

Consolidated Appropriations Act, 2008. For Rural Utilities Service, U.S. Department of Agriculture, provides $6.45 million to support a loan level of $300 million for the broadband loan program, and $13.5 million for broadband community connect grants. Signed by President, December 26, 2007.

P.L. 110-234 (H.R. 2419)

Food, Conservation, and Energy Act of 2008. Reauthorizes broadband program at the Rural Utilities Service through FY2012. Establishes a National Center for Rural Telecommunications Assessment. Directs USDA and the FCC to submit to Congress a comprehensive rural broadband strategy. Introduced May 22, 2007; referred to Committee on Agriculture, and in addition to Committee on Foreign Affairs. Subcommittee on Specialty Crops, Rural Development, and Foreign Agriculture held markup of Title VII (Rural Development) on June 6, 2007. Reported by House Committee on Agriculture (H.Rept. 110-256) on July 23, 2007. Passed House July 27, 2007. Passed Senate with an amendment, December 14, 2007. Conference report (H.Rept. 110-627) approved by the House May 14, 2008, and by the Senate May 15, 2008. Vetoed by the President, May 21, 2008. House and Senate overrode veto on May 21 and May 22, 2008. Became Public Law 110-234.

H.R. 42 (Velazquez)

Serving Everyone with Reliable, Vital Internet, Communications and Education Act of 2007. Directs the FCC to expand assistance provided by the Lifeline Assistance Program and the Link Up Program to include broadband service. Introduced January 4, 2007; referred to Committee on Energy and Commerce.

H.R. 1818 (Matsui)

Broadband Deployment Acceleration Act of 2007. Amends the Internal Revenue Code of 1986 to provide for the expensing of broadband Internet access expenditures. Introduced March 29, 2007; referred to Committee on Ways and Means.

H.R. 2035 (Herseth Sandlin)

Rural Broadband Improvement Act. Amends the Rural Electrification Act of 1936 to modify the broadband loan program at the Rural Utilities Service by narrowing the definition of "eligible rural community" and by limiting loans awarded to applicants proposing to serve areas that already have a broadband provider. Introduced April 25, 2007; referred to Committee on Agriculture and to Committee on Energy and Commerce.

H.R. 2054 (Boucher)

Universal Reform Act of 2007. Targets universal service support specifically to eligible telecommunications carriers in high-cost geographic areas to ensure that communications services and high-speed broadband services are made available throughout all of the States of the United States in a fair and equitable manner. Introduced April 26, 2007; referred to Committee on Energy and Commerce.

H.R. 2174 (Salazar)

Rural Broadband Initiative Act of 2007. Establishes an Office of Rural Broadband Initiatives within the Department of Agriculture which will administer all rural broadband grant and loan programs previously administered by the Rural Utilities Service. Also establishes a National Rural Broadband Innovation Fund which would fund experimental and pilot rural broadband projects and applications. Introduced May 3, 2007; referred to Committee on Agriculture and to Committee on Energy and Commerce.

H.R. 2569 (Graves)

Rural Broadband Deployment Act. Codifies certain changes proposed by USDA to the rules governing eligibility for the rural broadband access program.

Specifically, would relax market survey requirements and eliminate the credit support requirement, including the cash-on-hand requirement. Introduced June 5, 2007; referred to Committee on Agriculture, and in addition to the Committee on Energy and Commerce.

H.R. 2720 (Kind)

FARM 21 Act of 2007. Amends the Farm Security and Rural Investment Act of 2002 to direct that the Secretary of USDA shall make available funds of the Commodity Credit Corporation to the rural broadband loan program as follows: $10 million for each of fiscal years 2008 through 2012. Also specifies criteria to be applied by USDA in considering applications for all rural development projects. Introduced June 14, 2007; referred to Committee on Agriculture, and in addition to the Committees on Education and Labor, Foreign Affairs, and Ways and Means.

H.R. 2953 (Space)

Rural Broadband Access Enhancement Act. Seeks to redefine “eligible rural community, streamline application process and lowers equity requirements, restricts loans to communities with existing broadband providers, eliminates limitation on eligibility based on number of subscriber lines, sets 35-year maximum on term of loan repayment, and directs USDA/RUS to meet specific reporting requirements. Introduced July 10, 2007; referred to Committee on Agriculture and Committee on Energy and Commerce.

H.R. 3161 (DeLauro)

Agriculture, Rural Development, Food and Drug Administration, and Related Agencies Appropriations, 2008. Provides \$6.45 million to support a loan level of \$300 million for the broadband loan program, and \$17.82 million for broadband community connect grants. Introduced July 24, 2007; referred to Committee on Appropriations. Reported by Committee on Appropriations, July 24, 2007 (H.Rept. 110-258; placed on Union Calendar. Passed House, August 2, 2007.

H.R. 3281 (Boucher)

Community Broadband Act of 2007. Sets forth that no state regulation or requirement shall prevent a public provider from offering broadband services, and prohibits a municipality from discriminating against competing private providers. Introduced August 1, 2007; referred to Committee on Energy and Commerce.

H.R. 3246 (Oberstar)

Regional Economic and Infrastructure Development Act of 2007. Designates five regional commissions throughout the U.S. which would provide economic and infrastructure development grants, including grants to develop the telecommunications infrastructure of the region. Introduced July 31, 2007; referred to Committee on Transportation and Infrastructure and to Committee on Financial Services. Reported by Committee on Transportation and Infrastructure, September 7, 2007 (H.Rept. 110-321, Part I). Passed House, October 4, 2007.

H.R. 3428 (McHugh)

Rural America Digital Accessibility Act. Provides for grants, loan guarantees, research, and tax credits to promote broadband deployment in underserved rural areas. Introduced August 3, 2007; referred to Committee on Energy and Commerce and in addition to the Committee on Ways and Means and the Committee on Science and Technology.

H.R. 3627 (Space)

Connect the Nation Act. Establishes a State Broadband Data and Development Grant Program within the Department of Commerce to help states develop and implement statewide initiatives to identify and track the availability and adoption of broadband services within each state. Authorizes $40 million for each of fiscal years 2008 through 2012. Introduced September 20, 2007; referred to Committee on Energy and Commerce.

H.R. 3893 (Allen)

Connect America Now Act. Establishes a State Broadband Data and Development Grant Program within the Department of Commerce to help states develop and implement statewide initiatives to identify and track the availability and adoption of broadband services within each state. Authorizes $40 million for each of fiscal years 2008 through 2012. Introduced October 18, 2007; referred to Committee on Energy and Commerce.

H.R. 3919 (Markey)

Broadband Census of America Act of 2007. Provides for a comprehensive inventory of existing broadband service. Directs the FCC to conduct an annual assessment of broadband deployment, including information on bandwidth service tiers, types of technology, and international comparisons. Directs NTIA to develop and maintain a broadband inventory map of the United States that depicts broadband deployment at a nine digit zip code area level, census tract level, or functional equivalent. Directs NTIA to award grants to states for broadband map development and grants for demand-side broadband service identification and assessments. Directs the FCC to conduct periodic consumer surveys of broadband service capability. Authorizes $20 million for each of fiscal years 2008 through 2010, of which not less than $15 million would be available for the state broadband map grants. Authorizes $50 million in FY2008, $100 million in FY2009, and $125 million in FY20 10 for the demand-side broadband service identification and assessment (local technology planning) grants. Introduced October 22, 2007; referred to Committee on Energy and Commerce. Reported by Committee on Energy and Commerce (H.Rept. 110-443), November 13, 2007. Passed House by voice vote, November 13, 2007.

H.R. 5682 (Allen)

Rural America Communication Expansion for the Future Act of 2008. Reforms and reauthorizes through FY2013 the Rural Broadband Access Loan and Loan Guarantee Program and the Community Connect Grant Program. Provides for tax incentives and NTIA grant program for broadband services in rural and underserved areas. Introduced April 2, 2008; referred to Committee on Energy and Commerce and in addition to the Committees on Ways and Means and Agriculture.

H.R. 6320 (Markey)

Twenty-first Century Communications and Video Accessibility Act of 2008. Ensures that individuals with disabilities have access to emerging Internet Protocol-based communication and video programming technologies in the 21st Century. Introduced June 19, 2008; referred to Committee on Energy and Commerce.

H.R. 6356 (Barton)

Universal Service Reform, Accountability, and Efficiency Act of 2008. Reforms the collection and distribution on universal service support under the Communications Act of 1934. Introduced June 24, 2008; referred to Committee on Energy and Commerce.

H.Res. 1292 (Eshoo)

Establishes a national goal for the universal deployment of next-generation broadband networks by 2015 and calls upon the Congress and the President to develop a strategy, enact legislation, and adopt policies to accomplish this objective. Introduced June 20, 2008; referred to Committee on Energy and Commerce.

S. 101 (Stevens)

Universal Service for Americans Act ("USA Act"). Directs the FCC to establish Broadband for Unserved Area Areas Program to be funded by the Universal Service Fund. Requires communications carriers to submit detailed broadband deployment data to the FCC. Introduced January 4, 2007; referred to Committee on Commerce, Science, and Transportation.

S. 541 (Feingold)

Rural Opportunities Act of 2007. Directs the FCC to collect more detailed broadband deployment data and to periodically revise its definition of broadband above 200 kbps. Directs the Secretary of Agriculture to report on the adoption or planned adoption of the recommendations contained in the September 2005 audit report by the Inspector General of the United States Department of Agriculture. Introduced February 8, 2007; referred to Committee on Agriculture, Nutrition and Forestry.

S. 711 (Smith)

Universal Service for the 21st Century Act. Expands the contribution base for universal service and establishes a separate account within the universal service fund to support the deployment of broadband service in unserved areas. Introduced February 28, 2007; referred to Committee on Commerce, Science, and Transportation.

S. 761 (Reid)

America COMPETES Act. Authorizes the National Science Foundation (NSF) to provide grants for basic research in advanced information and communications technologies. Areas of research include affordable broadband access, including wireless technologies. Also directs NSF to develop a plan that describes the current status of broadband access for scientific research purposes. Introduced March 5, 2007; placed on Senate Legislative Calendar, March 6, 2007. Passed Senate April 25, 2007. Senate incorporated this measure in H.R. 2272 as an amendment July 19, 2007.

S. 1032 (Clinton)

Rural Broadband Initiative Act of 2007. Establishes an Office of Rural Broadband Initiatives within the Department of Agriculture which will administer all rural broadband grant and loan programs previously administered by the Rural Utilities Service. Also establishes a National Rural Broadband Innovation Fund which would fund experimental and pilot rural broadband projects and applications. Introduced March 29, 2007; referred to Committee on Agriculture, Nutrition, and Forestry.

S. 1190 (Durbin)

Connect the Nation Act. Establishes a State Broadband Data and Development Grant Program within the Department of Commerce to help states develop and implement statewide initiatives to identify and track the availability and adoption of broadband services within each state. Authorizes $40 million for each of fiscal years 2008 through 2012. Introduced April 24, 2007; referred to Committee on Commerce, Science, and Transportation.

S.Res. 191 (Rockefeller)

Establishing a national goal for the universal deployment of next-generation broadband networks by 2015, and calling upon Congress and the President to develop a strategy, enact legislation, and adopt policies to accomplish this objective. Introduced May 8, 2007; referred to Committee on Commerce, Science, and Transportation.

S. 1264 (Coleman)

Rural Renaissance Act. Creates a Rural Renaissance Corporation which would fund qualified projects including projects to expand broadband technology in rural areas. Introduced May 2, 2007; referred to Committee on Finance.

S. 1439 (Roberts)

Rural Broadband Improvement Act of 2007. Reauthorizes the broadband and broadband loan guarantee program under Title VI of the Rural Electrification Act of 1936. Introduced May 21, 2007; referred to Committee on Agriculture, Nutrition, and Forestry.

S. 1492 (Inouye)

Broadband Data Improvement Act. Seeks to improve the quality of federal broadband data collection and encourage state initiatives that promote broadband deployment. Directs the FCC to reevaluate its current 200 kbps broadband standard and to develop a new metric for "second generation broadband" capable of transmitting high definition video content. Directs broadband providers to report to the FCC connections within either census tracts, nine digit (zip+4) postal zip code areas, or five digit postal zip code areas. Directs the FCC to conduct its Section 706 inquiry into the status of broadband deployment on an annual basis. Directs the Census Bureau to collect residential broadband data. Directs GAO to develop broadband metrics involving connection cost and capability information that could be used to improve the process of comparing U.S. broadband deployment with other countries. Directs the Small Business Administration to conduct a study evaluating the impact of broadband speed and cost on small businesses. Authorizes $40 million for each of fiscal years 2008 through 2012 to establish a State Broadband Data and Development Grant Program within the Department of Commerce to help states develop and implement statewide initiatives to identify and track the availability and adoption of broadband services within each state. Introduced May 24, 2007; referred to Committee on Commerce, Science, and Transportation. Ordered to be reported July 19, 2007; reported by Committee (S.Rept. 110-204) and placed on Senate Legislative Calendar, October 24, 2007.

S. 1745 (Mikulski)

Departments of Commerce and Justice, Science, and Related Agencies Appropriations Act, 2008. Provides $10 million to the Technology Opportunities Program (TOP) at the National Telecommunications and Information Administration, Department of Commerce, for competitive grants for the construction of broadband services. The Senate Appropriations Committee expects NTIA to give preference to rural projects promoting broadband deployment in support of educational, cultural, healthcare, or other social services. Introduced June 29, 2007; referred to Committee on Appropriations. Reported to Senate, June 29, 2007

(S.Rept. 110-124); placed on Senate Legislative Calendar. Agreed to by Senate in the nature of a substitute (as *H.R. 3093*), October 4, 2007. Passed Senate October 16, 2007.

S. 1853 (Lautenberg)

Community Broadband Act of 2007. Sets forth that no state regulation or requirement shall prevent a public provider from offering broadband services, and prohibits a municipality from discriminating against competing private providers. Introduced July 23, 2007; referred to Committee on Commerce, Science, and Transportation. Reported with amendments (S.Rept. 110-330) by the Committee on Commerce, Science and Transportation, April 22, 2008.

S. 1859 (Kohl)

Agriculture, Rural Development, Food and Drug Administration, and Related Agencies Appropriations, 2008. Provides $ 10.643 million to support a loan level of $495 million for the broadband loan program, and $8.9 million for broadband grants. Introduced July 24, 2007; referred to Committee on Appropriations. Reported to Senate (S.Rept. 110-134) July 24, 2007; placed on Senate Legislative Calendar.

S. 2242 (Baucus)

Heartland, Habitat, Harvest, and Horticulture Act of 2007. Introduced October 25, 2007; referred to Committee on Finance. Amends the Internal Revenue Code of 1986 to provide for the expensing of broadband Internet access expenditures. Creates a Rural Renaissance Corporation which would fund qualified projects including projects to expand broadband technology in rural areas. Reported to Senate (S.Rept. 110-206) and placed on Senate Legislative Calendar, October 25, 2007.

S. 2302 (Harkin)

Food and Energy Security Act of 2007. Reauthorizes broadband program at the Rural Utilities Service through FY20 12. Introduced November 2, 2007. Senate Committee on Agriculture, Nutrition, and Forestry reported measure to Senate (S.Rept. 110-220) November 2, 2007; placed on Senate Legislative Calendar.

S. 3182 (Mikulski)

Departments of Commerce and Justice, Science, and Related Agencies Appropriations Act, 2009. Provides $20 million to the Technology Opportunities Program (TOP) at the

National Telecommunications and Information Administration, Department of Commerce, for competitive grants for the construction of broadband services and networks. Reported by Committee on Appropriations June 23, 2008 (S.Rept. 110-397).

S. 3260 (Durbin)

Financial Services and General Government Appropriations for FY2009. Makes $3 million available to the FCC to establish and administer a State Broadband Data and Development matching grants program for State-level broadband demand aggregation activities and creation of geographic inventory maps of broadband service to identify gaps in service and provide a baseline assessment of statewide broadband deployment. Reported by Committee on Appropriations July 14, 2008 (S.Rept. 110-417).

S. 3289 (Kohl)

Agriculture, Rural Development, Food and Drug Administration, and Related Agencies Appropriations, 2009. Provides $1 1.618 million to support a loan level of $297.9 million for the broadband loan program, and $ 13.406 million for broadband grants. Introduced July 21, 2008; referred to Committee on Appropriations. Reported to Senate (S.Rept. 110-426) July 21, 2008; placed on Senate Legislative Calendar.

S. 3297 (Reid)

Advancing America's Priorities Act. Title V, Subtitle A, Part 1 is the Broadband Data Improvement Act, which seeks to improve the quality of federal broadband data collection and encourage state initiatives that promote broadband deployment. Introduced July 22, 2008; placed on Senate Legislative Calendar.

POLICY ISSUES

Legislation in the 110th Congress seeks to provide federal financial assistance for broadband deployment in rural and underserved areas. In assessing this legislation, several policy issues arise.

Is Broadband Deployment Data Adequate?

Obtaining an accurate snapshot of the status of broadband deployment is problematic. Anecdotes abound of rural and low-income areas which do not have adequate Internet access, as well as those which are receiving access to high-speed, state-of-the-art connections. Rapidly evolving technologies, the constant flux of the telecommunications industry, the uncertainty of consumer wants and needs, and the sheer diversity and size of the nation's economy and geography make the status of broadband deployment very difficult to characterize. The FCC periodically collects broadband deployment data from the private sector via "FCC Form 477" — a standardized information gathering survey. Statistics derived from the Form 477 survey are published every six months. Additionally, data from Form 477 are used as the basis of the FCC's (to date) five broadband deployment reports.

The FCC is working to refine the data used in future Reports in order to provide an increasingly accurate portrayal. In its March 17, 2004 Notice of Inquiry for the *Fourth Report*, the FCC sought comments on specific proposals to improve the FCC Form 477 data gathering program.[55] On November 9, 2004, the FCC voted to expand its data collection program by requiring reports from all facilities based carriers regardless of size in order to better track rural and underserved markets, by requiring broadband providers to provide more information on the speed and nature of their service, and by establishing broadband-over-power line as a separate category in order to track its development and deployment. The FCC Form 477 data gathering program was extended for five years beyond its March 2005 expiration date.[56]

The Government Accountability Office (GAO) has cited concerns about the FCC's zip-code level data. Of particular concern is that the FCC will report broadband service in a zip code even if a company reports service to only one subscriber, which in turn can lead to some observers overstating broadband deployment. According to GAO, "the data may not provide a highly accurate depiction of local deployment of broadband infrastructures for residential service, especially in rural areas." The FCC has acknowledged the limitations in its zip code level data.[57]

On April 16, 2007, the FCC announced a Notice of Proposed Rulemaking which seeks comment on a number of broadband data collection issues, including how to develop a more accurate picture of broadband deployment; gathering information on price, other factors determining consumer uptake of broadband, and international comparisons; how to improve data on wireless broadband; how to collect information on subscribership to voice over Internet Protocol service (VoIP); and whether to modify collection of speed tier information.[58]

On March 19, 2008, the FCC adopted an Order that substantially expands its broadband data collection capability. Specifically, the Order expands the number of broadband reporting speed tiers to capture more information about upload and download speeds offered in the marketplace, requires broadband providers to report numbers of broadband subscribers by census tract, and improves the accuracy of information collected on mobile wireless broadband deployment. Additionally, in a Further Notice of Proposed Rulemaking, the FCC is seeking comment on broadband service pricing and availability.[59]

State initiatives to collect broadband deployment data in order to promote broadband in underserved areas are being viewed by Congress as a possible model for governmental efforts to encourage broadband. In particular, an initiative in the Commonwealth of Kentucky — called ConnectKentucky — has developed detailed broadband inventory mapping which identifies local communities that lack adequate broadband service. Kentucky is using this data to promote public-private partnerships in order to reach a goal of universal broadband coverage in the state.[60] Other states are pursuing or considering similar approaches.

On the federal level, Congress is exploring ways to support or implement the types of broadband mapping and data collection efforts demonstrated by ConnectKentucky. H.R.

3919, the Broadband Census of America Act of 2007, was passed by the House on November 13, 2007. H.R. 3919 would: (1) direct the FCC to conduct an annual assessment of broadband deployment, including information on bandwidth service tiers, types of technology, and international comparisons; (2) direct NTIA to develop and maintain a broadband inventory map of the United States that depicts broadband deployment at a nine digit zip code area level, census tract level, or functional equivalent; (3) direct NTIA to award grants to states for broadband map development and grants for demand-side broadband service identification and assessments; and (4) direct the FCC to conduct periodic consumer surveys of broadband service capability. H.R. 3919 would authorize $20 million for each of fiscal years 2008 through 2010, of which not less than $15 million would be available for the state broadband map grants. Additionally, H.R. 3919 would authorize $50 million in FY2008, $100 million in FY2009, and $125 million in FY20 10 for the demand-side broadband service identification and assessment (local technology planning) grants.

In the Senate, S. 1492, the Broadband Data Improvement Act introduced by Senator Inouye on May 24, 2007, and reported by the Senate Committee on Commerce, Science, and Transportation on July 19, 2007 (S.Rept. 110-204), seeks to improve the quality of federal broadband data collection and encourage state initiatives that promote broadband deployment. Specifically, S. 1492 would: (1) direct the FCC to reevaluate its current 200 kbps broadband standard and to develop a new metric for "second generation broadband" capable of transmitting high definition video content; (2) direct broadband providers to report to the FCC connections within either census tracts, nine digit (zip+4) postal zip code areas, or five digit postal zip code areas; (3) direct the FCC to conduct its Section 706 inquiry into the status of broadband deployment on an annual basis; (4) direct the Census Bureau to collect residential broadband data; direct GAO to develop broadband metrics involving connection cost and capability information that could be used to improve the process of comparing U.S. broadband deployment with other countries; (5) direct the Small Business Administration to conduct a study evaluating the impact of broadband speed and cost on small businesses; and (6) authorize $40 million for each of fiscal years 2008 through 2012 to establish a State Broadband Data and Development Grant Program within the Department of Commerce to help states develop and implement statewide initiatives to identify and track the availability and adoption of broadband services within each state. The provision establishing state broadband data grants matches language in the Connect the Nation Act (S. 1190), previously introduced by Senator Durbin on April 24, 2007.

Meanwhile, the Senate farm bill, H.R. 2419 as passed by the Senate on December 14, 2007, also contained a "Connect the Nation Act" subtitle. However, this provision was dropped during the farm bill conference.

Is Federal Assistance for Broadband Deployment Premature or Inappropriate?

Related to the data issue is the argument that government intervention in the broadband marketplace would be premature or inappropriate. Some argue that financial assistance for broadband deployment could distort private sector investment decisions in a dynamic and rapidly evolving marketplace, and question whether federal tax dollars should support a technology that has not yet matured, and whose societal benefits have not yet been demonstrated.[61]

On the other hand, proponents of financial assistance counter that the available data show, in general, that the private sector will invest in areas where it expects the greatest return: areas of high population density and income. Without some governmental assistance in underserved areas, they argue, it is reasonable to conclude that broadband deployment will lag behind in many rural and low income areas.[62]

Which Approach Is Best?

If one assumes that governmental action is appropriate to spur broadband deployment in underserved areas, which specific approaches, either separately or in combination, would likely be most effective? Targeted grants and loans from several existing federal programs have been proposed, as well as tax credits for companies deploying broadband systems in rural and low-income areas. Another approach is spectrum policy to encourage wireless broadband deployment.[63] How might the impact of federal assistance compare with the effects of regulatory or deregulatory actions?[64] And finally, how might any federal assistance programs best compliment existing "digital divide" initiatives by the states, localities, and private sector?

Table 1. Selected Federal Domestic Assistance Programs Related to Telecommunications Development

Program	Agency	Description	FY2008 (obligations)	Web Links for More Information [http://12.46.245.173/cfda/cfda.html]: Go to "All Programs Listed Numerically" and search by program
Public Telecommunications Facilities — Planning and Construction	National Telecommunications and Information Administration, Dept. of Commerce	Assists in planning, acquisition, installation and modernization of public telecommunications facilities	$19.5 million	[http://www.ntia.doc.gov/otiahome/ptfp/index.html]
Investments for Public Works and Economic Development Facilities	Economic Development Administration, Dept. of Commerce	Provides grants to economically distressed areas for construction of public facilities and infrastructure, including broadband deployment and other types of telecommunications enabling projects	$249 million	[http://www.eda.gov/]

Rural Telephone Loans and Loan Guarantees	Rural Utilities Service, U.S. Dept. of Agriculture	Provides long-term direct and guaranteed loans to qualified organizations for the purpose of financing the improvement, expansion, construction, acquisition, and operation of telephone lines, facilities, or systems to furnish and improve telecommunications service in rural areas	$145 million (hardship loans); $250 million (cost of money loans); $295 million (FFB Treasury loans)	[http://www.usda.gov/rus/telecom/index.htm]
Distance Learning and Telemedicine Loans and Grants	Rural Utilities Service, U.S. Dept. of Agriculture	Provides seed money for loans and grants to rural community facilities (e.g., schools, libraries, hospitals) for advanced telecommunications systems that can provide health care and educational benefits to rural areas	$24.7 million (grants) $128 million (loans) (FY2007)	[http://www.usda.gov/rus/telecom/dlt/dlt.htm]
Rural Broadband Access Loan and Loan Guarantee Program	Rural Utilities Service, U.S. Dept. of Agriculture	Provides loan and loan guarantees for facilities and equipment providing broadband service in rural communities	$300 million (cost of money loans)	[http://www.usda.gov/rus/telecom/broadband.htm]
Community Connect Broadband Grants	Rural Utilities Service, U.S. Dept. of Agriculture	Provides grants to applicants proposing to provide broadband service on a "community-oriented connectivity" basis to rural communities of under 20,000 inhabitants.	$13.4 million	[http://www.usda.gov/rus/telecom/index.htm]
Education Technology State Grants	Office of Elementary and Secondary Education, Dept. of Education	Grants to State Education Agencies for development of information technology to improve teaching and learning in schools	$267 million	[http://www.ed.gov/Technology/TLCF/index.html]
Ready to Teach	Office of Assistant Secretary for Educational Research and Improvement, Dept. of Education	Grants to carry out a national telecommunication-based program to improve the teaching in core curriculum areas.	$10.7 million	[http://www.ed.gov/programs/readyteach/index.html]

Table 1. Continued

Special Education — Technology and Media Services for Individuals with Disabilities	Office of Special Education and Rehabilitative Services, Dept. of Education	Supports development and application of technology and education media activities for disabled children and adults	$39.3 million	[http://www.ed.gov/about/offices/list/osers/index.html?src=mr/]
Telehealth Network Grants	Health Resources and Services Administration, Department of Health and Human Services	Grants to develop sustainable telehealth programs and networks in rural and frontier areas, and in medically unserved areas and populations.	$3.9 million	[http://www.hrsa.gov/telehealth/]
Medical Library Assistance	National Library of Medicine, National Institutes of Health, Department of Health and Human Services	Provides funds to train professional personnel; strengthen library and information services; facilitate access to and delivery of health science information; plan and develop advanced information networks; support certain kinds of biomedical publications; and conduct research in medical informatics and related sciences	$67.5 million	[http://www.nlm.nih.gov/ep/extramural.html]
State Library Program	Office of Library Services, Institute of Museum and Library Services, National Foundation on the Arts and the Humanities	Grants to state library administrative agencies for promotion of library services that provide all users access to information through State, regional, and international electronic networks	$171.5 million	[http://www.imls.gov/grants/library/lib_gsla.asp#po]
Native American and Native Hawaiian Library Services	Office of Library Services, Institute of Museum and Library Services, National Foundation on the Arts and the Humanities	Supports library services including electronically linking libraries to networks	$3.7 million	[http://www.imls.gov/grants/library/lib_nat.asp]
Appalachian Area Development	Appalachian Regional Commission	Provides project grants for Appalachian communities to support the physical	$73 million	[http://www.arc.gov/index.do?nodeId=21]

		infrastructure necessary for economic development and improved quality of life.		
Delta Area Economic Development	Delta Regional Authority	Grants to support self-sustaining economic development of eight states in Mississippi Delta region.	$7.8 million	[http://www.dra.gov/programs/information-technology]
Denali Commission Program	Denali Commission	Provides grants through a federal and state partnership designed to provide critical infrastructure and utilities throughout Alaska, particularly in distressed communities	$106 million	[http://www.denali.gov/]

Source: Prepared by CRS based on information from the Catalog of Federal Domestic Assistance.

Table 2. Selected Federal Programs Funding Broadband Access

Program	Comments
Programs Funding Access to Telecommunications in Underserved Areas	
Rural Broadband Access Loan and Loan Guarantee Program (Rural Utilities Service, U.S. Department of Agriculture)	Provides loan and loan guarantees for facilities and equipment providing broadband service in rural communities.
Community Connect Broadband Grants (Rural Utilities Service, U.S. Department of Agriculture)	Provides grants to applicants proposing to provide broadband service on a "community-oriented connectivity" basis to rural communities of under 20,000 inhabitants.
Rural Telephone Loans and Loan Guarantees (Rural Utilities Service, U.S. Department of Agriculture)	Since 1995, the RUS Rural Telephone Loan and Loan Guarantee program — which has traditionally financed telephone voice service in rural areas under 5,000 inhabitants — has required that all telephone facilities receiving financing must be capable of providing DSL broadband service at a rate of at least 1 megabyte per second.
Universal Service Fund: High Cost Program (Federal Communications Commission)	While the USF's High Cost Program does not explicitly fund broadband infrastructure, subsidies are used, in many cases, to upgrade existing telephone networks.
Federal Economic Development Programs Funding Broadband Access	
Community Development Block Grants (Department of Housing and Urban Development)	In Michigan, a Digital Divide Investment Program (DDIP) combined Michigan Broadband Development Authority loans (initially $12 million) and CDBG grant funding ($4 million) to deploy a hybrid fixed wireless and fiber network in two rural counties which would make broadband affordable for low to moderate income residents.

Table 2. Continued

Indian Community Development Block Grants (Department of Housing and Urban Development)	In 2005, the HUD awarded the Coquille Indian Tribe a $421,354 grant used to fund the Coquille Broadband Technology Infrastructure Project. The project will allow for improved connectivity for reservation residents, improvements in rural community access, and potentially increased wireless Internet access for the Tribal and surrounding communities.
Grants for Public Works and Economic Development Facilities (Economic Development Administration, Department of Commerce)	Supports the proliferation of broadband networks as a key priority for regional economic growth. Examples: $6 million grant to a company in Virginia for investment in 300 miles of fiber optic cable in nine counties and three cities; $2 million grant to companies in Vermont to help build a 424 mile fiber optic broadband network in rural northern Vermont; and $270 thousand to support a Rhode Island Wireless Innovation Networks project. EDA encourages communities eligible for RUS programs to access that first before applying for EDA investment dollars.
Appalachian Regional Commission	The Appalachian Regional Development Act Amendments of 2002 reauthorized ARC for five years and created specific authority for a Region-wide initiative to bridge the telecommunications and technology gap between the Appalachian Region and the rest of the United States. Supported a telecommunications initiative ($33 million over five year period) which includes projects such as: a regional fiber network across northeast Mississippi; wireless demonstrations in rural New York, Ohio, Pennsylvania, Virginia, West Virginia, and Georgia; and a regionwide effort in Kentucky to compile an inventory of broadband access across the 51 Appalachian counties and work with the private sector to substantially increase broadband coverage. In Maryland, a county-wide high-speed wireless network, funded by ARC over several years, now serves over 4,500 customers.
Delta Regional Authority	During a strategic planning retreat in February 2005, the DRA board determined that one of the authority's three top policy priorities would be information technology. To support its policy position, the authority devoted $150,000 to create an information technology plan for the region.
Denali Commission	Funded Telecommunications Survey in 2000 which was used to determine the state of broadband deployment in Alaska and used as basis for applying for RUS broadband assistance.
Applications-Based Federal Programs Related to Broadband	
Universal Service Fund: Schools and Libraries or "E-Rate" Program (Federal Communications Commission)	Used to fund broadband access for schools and libraries.

Universal Service Fund: Rural Health Care Program (Federal Communications Commission)	Used to fund broadband access for rural health care centers.
Distance Learning and Telemedicine Program (Rural Utilities Service, U.S. Department of Agriculture)	Provides seed money for loans and grants to rural community facilities (e.g., schools, libraries, hospitals) for advanced telecommunications systems that can provide health care and educational benefits to rural areas.
Public Safety Interoperable Communications Grant Program (National Telecommunications and Information Administration, Department of Commerce)	Provides funding to states and territories to enable and enhance public safety agencies' interoperable communications capabilities.
Telehealth Network Grants (Health Resources and Services Administration, Department of Health and Human Services)	Grants to develop sustainable telehealth programs and networks in rural and frontier areas, and in medically unserved areas and populations.
Public Telecommunications Facilities Program (National Telecommunications and Information Administration, Department of Commerce)	Grants for public television, public radio, and nonbroadcast distance learning projects.
Education technology programs (Department of Education)	Examples include Education Technology State Grants, Ready to Teach.
State Library Grants (Office of Library Services, Institute of Museum and Library Services, National Foundation on the Arts and the Humanities)	Grants to state library administrative agencies for promotion of library services that provide all users access to information through State, regional, and international electronic networks.
Medical Library Assistance (National Library of Medicine, National Institutes of Health, Department of Health and Human Services)	Provides funds to train professional personnel; strengthen library and information services; facilitate access to and delivery of health science information; plan and develop advanced information networks; support certain kinds of biomedical publications; and conduct research in medical informatics and related sciences.

ENDNOTES

[1] The term "digital divide" can also refer to international disparities in access to information technology. This report focuses on domestic issues only.

[2] See U.S. Department of Commerce, *Falling Through the Net: Toward Digital Inclusion*, released October 2000.

[3] Not all observers agree that a "digital divide" exists. See, for example: Thierer, Adam D., *Divided Over the Digital Divide*, Heritage Foundation, March 1, 2000. [http://www.heritage.org/Press/Commentary/ED030100.cfm]

[4] Department of Commerce, *A Nation Online: How Americans Are Expanding Their Use of the Internet*, February 2002. Based on a September 2001 Census Bureau survey of 57,000 households. See [http://www.ntia.doc.gov/ntiahome/dn/index.html].

[5] *A Nation Online*, pp. 10-11.

[6] For further information on different types of broadband technologies, including their respective strengths and limitations, see CRS Report RL33542, *Broadband Internet Access: Background and Issues*, by Angele A. Gilroy and Lennard G. Kruger.

[7] FCC, *High-Speed Services for Internet Access: Status as of June 30, 2007,* March 2008. Available at [http://hraunfoss.fcc.gov/edocs_public/attachmatch/DOC-280906A1 .pdf].

[8] Percentage assumes one high speed line per household, 65.9 million residential high speed lines (per June 30, 2007 FCC data) and 114 million households in the United States (2006 Census data, see [http://www.census.gov/prod/2007pubs/08abstract/pop.pdf]).

[9] Federal Communications Commission, *Fourth Report to Congress*, "Availability of Advanced Telecommunications Capability in the United States," GN Docket No. 04-54, FCC 04-208, September 9, 2004, p. 38. Available at [http://hraunfoss.fcc.gov/edocs_public/attachmatch/FCC-04-208A1 .pdf].

[10] For more information on rural broadband and broadband programs at the Rural Utilities Service, see CRS Report RL33816, *Broadband Loan and Grant Programs in the USDA's Rural Utilities Service, by* Lennard G. Kruger.

[11] See U.S. Depts. of Commerce and Agriculture, *Advanced Telecommunications in Rural America: The Challenge of Bringing Broadband Service to All Americans*, April 2000, 80 pp. Available at [http://www.ntia.doc.gov/reports/ruralbb42600.pdf]

[12] U.S. Government Accountability Office, *Broadband Deployment is Extensive throughout the United States, but It Is Difficult to Assess the Extent of Deployment Gaps in Rural Areas*, GAO-06-426, May 2006, p. 12.

[13] Horrigan, John B., Pew Internet & American Life Project, *Home Broadband Adoption 2008,* July 2008, p. 3. Available at [http://www.pewinternet.org/pdfs/PIP_Broadband_2008.pdf].

[14] FCC, *High-Speed Services for Internet Access: Status as of June 30, 2007,* p. 4.

[15] U.S. Department of Commerce, National Telecommunications and Information Administration, *Networked Nation: Broadband in America 2007*, January 2008, 60 pp.

[16] NTIA, Households using the Internet in and outside the home, by selected characteristics, October 2007. Available at [http: //www.ntia.doc .gov/reports/2008/Table_HouseholdInternet2007.pdf].

[17] Horrigan, John B., Pew Internet & American Life Project, *Home Broadband Adoption 2008,* July 2008, p. 3. Available at [http://www.pewinternet.org/pdfs/PIP_Broadband_2008.pdf].

[18] Crandall, Robert W. et al, *The Effect of Ubiquitous Broadband Adoption on Investment, Jobs, and the U.S. Economy*, Conducted by Criterion Economics, L.L.C. for the New Millennium Research Council, September 2003. Available at [http://www.newmillennium research.org/archive/bbstudyreport_091703.pdf]

[19] Gillett, Sharon E., Massachusetts Institute of Technology, *Measuring Broadband's Economic Impact*, report prepared for the Economic Development Administration, U.S. Department of Commerce, February 28, 2006 p. 4.

[20] Crandall, Robert, William Lehr, and Robert Litan, *The Effects of Broadband Deployment on Output and Employment: A Cross-sectional Analysis of U.S. Data*, June 2007, 20 pp. Available at [http://www3 .brookings.edu/views/papers/crandall/200706litan.pdf].

[21] International Telecommunications Union, *Economies by broadband penetration, 2007.* Available at [http://www.itu.int/ITU-D/ict/statistics/at_glance/top20_broad_2007.html].

[22] OECD, *OECD Broadband Statistics, December 2007.* Available at [http://www. oecd.org/sti/ict/broadband].

[23] OECD, Directorate for Science, Technology and Industry, *The Development of Broadband Access in OECD Countries*, October 29, 2001, 63 pp. For a comparison of government broadband policies, also see OECD, Directorate for Science, Technology and Industry, *Broadband Infrastructure Deployment: The Role of Government Assistance*, May 22, 2002, 42 pp.

[24] See Turner, Derek S., Free Press, *Broadband Reality Check II: The Truth Behind America's Digital Divide*, August 2006, pp 8-11. Available at [http://www.freepress.net/files/bbrc2-final.pdf]; and Turner, Derek S., Free Press, *'Shooting the Messenger' Myth vs. Reality: U.S. Broadband Policy and International Broadband Rankings*, July 2007, 25 pp., available at [http://www.freepress.net/files/shooting_the_messenger.pdf].

[25] National Telecommunications and Information Administration, *Fact Sheet: United States Maintains Information and Communication Technology (ICT) Leadership and Economic Strength*, at [http://www.ntia.doc.gov/ntiahome/press/2007/ICTleader_042407.html].

[26] See Wallsten, Scott, Progress and Freedom Foundation, *Towards Effective U.S. Broadband Policies*, May 2007, 19 pp. Available at [http://www.pff.org/issues-pubs/pops/pop14.7usbroadbandpolicy.pdf]. Also see Ford, George, Phoenix Center, *The Broadband Performance Index: What Really Drives Broadband Adoption Across the OECD?*, Phoenix Center Policy Paper Number 33, May 2008, 27 pp; available at [http://www.phoenixcenter.org/pcpp/PCPP33Final.pdf].

[27] See price and services and speed data on OECD Broadband Portal, available at [http://www.oecd.org/sti/ict/broadband]; Turner, Derek S., Free Press, *Broadband Reality Check II: The Truth Behind America's Digital Divide*, August 2006, pp 5-9; Kende, Michael, Analysis Consulting Limited, *Survey of International Broadband Offerings*, October 4, 2006, 12 p, available at [http://www.analysys.com/pdfs/Broadband Performance Survey.pdf]; and Atkinson, Robert D., The International Technology and Innovation Foundation, *Explaining International Broadband Leadership*, May 2008, 108 p, available at [http://www.itif.org/files/ExplainingBBLeadership.pdf].

[28] FCC News Release, "FCC Issues Report on the Deployment of Advanced Telecommunications Capability to All Americans," January 28, 1999. [http://www. fcc.gov/ Bureaus/Common _Carrier/News _Releases/1 999/nrcc9004.html]

[29] *Deployment of Advanced Telecommunications Capability: Second Report*, p. 6.

[30] *Third Report*, p. 5.

[31] Ibid., pp. 5-6.

[32] *Fourth Report*, p. 8.

[33] Ibid., p. 9.

[34] Ibid., pp. 5, 7.

[35] Federal Communications Commission, *Fifth Report*, "In the Matter of Inquiry Concerning the Deployment of Advanced Telecommunications Capability to All Americans in a Reasonable and Timely Fashion, and Possible Steps to Accelerate Such Deployment

Pursuant to Section 706 of the Telecommunications Act of 1996," GN Docket No. 07-45, FCC 08-88, Adopted March 19, 2008, Released June 12, 2008. 76 pp. Available at [http://hraunfoss.fcc.gov/edocs_public/attachmatch/FCC-08-88A1 .pdf].

[36] See speech by Nancy Victory, Assistant Secretary for Communications and Information, before the National Summit on Broadband Deployment, October 25, 2001, available at [http://www.ntia.doc.gov/speeches/200 1/broadband_1 02501 .htm].

[37] Address by Nancy Victory, NTIA Administrator, before the Alliance for Public Technology Broadband Symposium, February 8, 2002, [http://list.uvm.edu/cgi-bin/wa?A2 =ind0202B&L=COMMUNET&D=0&T=0&P=1 272&F=P].

[38] See White House, *A New Generation of American Innovation*, April 2004. Available at [http://www.whitehouse.gov/infocus/technology/economic_policy200404/innovation.pdf.

[39] Available at [http://www.technology.gov/reports/TechPolicy/Broadband_02092 1 .pdf].

[40] President's Council of Advisors on Science and Technology, Office of Science and Technology Policy, *Building Out Broadband*, December 2002, 14 pp. Available at [http://www.ostp.gov/pdf/final_broadband_report_with_letters.pdf].

[41] Available at [http://www.ntia.doc.gov/reports/2008/NetworkedNationBroadbandin America2007.pdf]

[42] NTIA, *Press Release*, "Gutierrez Hails Dramatic U.S. Broadband Growth," January 31, 2008. Available at [http://www.ntia.doc.gov/ntiahome/press/2008/NetworkedNation_0 1 3 108.html]

[43] California Broadband Task Force, *The State of Connectivity: Building Innovation Through Broadband*, Final Report of the California Broadband Task Force, January 2008, 83 pp. Available at [http://www.calink.ca.gov/pdf/CBTF _FINAL _Report.pdf].

[44] Wallsten, Scott, AEI-Brookings Joint Center for Regulatory Studies, *Broadband Penetration: An Empirical Analysis of State and Federal Policies*, Working Paper 05-12, June 2005, 29 pp. Available at [http://aei-brookings.org/admin/authorpdfs/page.php?id= 1161].

[45] For a summary of selected state broadband bills, see [http://www.ncsl.org/programs/lis/ legislation/BroadbandBills.htm]. For a listing of state broadband statutes, see [http://www.ncsl.org/programs/lis/legislation/broadbandstatutes.htm].

[46] NGA Center for Best Practices, *Issue Brief*, "State Efforts to Expand Broadband Access," May 20, 2008, 18 pp. Available at [http://www.nga.org/Files/pdf/0805BROADBAND ACCESS.PDF]

[47] The section on universal service was prepared by Angele Gilroy, Specialist in Telecommunications, Resources, Science and Industry Division. For more information on universal service, see CRS Report RL33979, *Universal Service Fund: Background and Options for Reform*, by Angele A. Gilroy.

[48] Communications Act of 1934, As Amended, Title I sec.1 [47 U.S.C. 151].

[49] Many states participate in or have programs that mirror FCC universal service mechanisms to help promote universal service goals within their states.

[50] Additional FCC policies such as rate averaging and pooling have also been implemented to assist high cost carriers.

[51] For additional information on this program, including funding commitments, see the E- rate website: [http://www.universalservice.org/sl/].

[52] For additional information on this program, including funding commitments, see the RHCD website: [http://www.universalservice.org/rhc/]

[53] For additional information on the TDF fund and TDF Foundation see the TDF website at [http://www.tdfund.com]

[54] The Joint Board recommended that the definition of those services that qualify for universal service support be expanded and that the nation's communications goals include the universal availability of: mobility services (i.e., wireless voice); broadband Internet services; and voice services at affordable and comparable rates for all rural and non-rural areas. For a copy of this recommendation see [http://hraunfoss.fcc.gov/edocs_public/attachmatch/FCC-07J-4A 1 .pdf].

[55] Federal Communications Commission, *Notice of Inquiry*, "Concerning the Deployment of Advanced Telecommunications Capability to All Americans in a Reasonable and Timely Fashion, and possible Steps to Accelerate Such Deployment Pursuant to Section 706 of the Telecommunications Act of 1996," FCC 04-55, March 17, 2004, p. 6.

[56] FCC News Release, *FCC Improves Data Collection to Monitor Nationwide Broadband Rollout*, November 9, 2004. Available at [http://hraunfoss.fcc.gov/edocs_public/attachmatch/DOC-2541 15A1 .pdf].

[57] U.S. Government Accountability Office, *Broadband Deployment is Extensive throughout the United States, but It Is Difficult to Assess the Extent of Deployment Gaps in Rural Areas*, GAO-06-426, May 2006, p. 3.

[58] Federal Communications Commission, *Notice Proposed Rulemaking*, "Development of Nationwide Broadband Data to Evaluate Reasonable and Timely Deployment of Advanced Services to All Americans, Improvement of Wireless Broadband Subscribership Data, and Development of Data on Interconnected Voice Over Internet Protocol (VoIP) Subscribership," WC Docket No. 07-38, FCC 07-17, released April 16, 2007, 56 pp.

[59] FCC, News Release, "FCC Expands, Improves Broadband Data Collection," March 19, 2008. Available at [http://hraunfoss.fcc.gov/edocs_public/attachmatch/DOC-280909A1 .pdf].

[60] Testimony of Brian Mefford, President and CEO, Connected Nation, Inc., before the Senate Committee on Commerce, Science and Transportation, April 24, 2007. Available at [http://commerce.senate.gov/public/_files/DC_Committeetestimony_04_23_07.pdf].

[61] See Leighton, Wayne A., *Broadband Deployment and the Digital Divide: A Primer*, a Cato Institute Policy Analysis, No. 410, August 7, 2001, 34 pp. Available at [http://www.cato.org/pubs/pas/pa410.pdf]. Also see Thierer, Adam, *Broadband Tax Credits, the High-Tech Pork Barrel Begins*, Cato Institute, July 13, 2001, available at [http://www.cato.org/tech/tk/0 1071 3-tk.html].

[62] See for example: Cooper, Mark, Consumer Federation of America and Consumers Union, *Expanding the Digital Divide & Falling Behind on Broadband*, October 2004, 33 pp. Available at [http://www.consumersunion.org/pub/ddnewbook.pdf].

[63] See CRS Report RS20993, *Wireless Technology and Spectrum Demand: Advanced Wireless Services*, by Linda K. Moore.

[64] See CRS Report RL33542, *Broadband Internet Access: Background and Issues*, by Angele A. Gilroy and Lennard G. Kruger, for a discussion of regulatory issues.

In: Maximizing Broadband Services to Rural Communities ISBN 978-1-60692-846-2
Editors: V.B. Haynesworth, A.R. Harrison, pp.69-100

Chapter 4

BROADBAND LOAN AND GRANT PROGRAMS IN THE USDA'S RURAL UTILITIES SERVICE*

Lennard G. Kruger

ABSTRACT

Given the large potential impact broadband access to the Internet may have on the economic development of rural America, concern has been raised over a "digital divide" between rural and urban or suburban areas with respect to broadband deployment. While there are many examples of rural communities with state of the art telecommunications facilities, recent surveys and studies have indicated that, in general, rural areas tend to lag behind urban and suburban areas in broadband deployment.

Citing the lagging deployment of broadband in many rural areas, Congress and the Administration acted in 2001 and 2002 to initiate pilot broadband loan and grant programs within the Rural Utilities Service (RUS) at the U.S. Department of Agriculture (USDA). Subsequently, Section 6103 of the Farm Security and Rural Investment Act of 2002 (P.L. 107-17 1) amended the Rural Electrification Act of 1936 to authorize a loan and loan guarantee program to provide funds for the costs of the construction, improvement, and acquisition of facilities and equipment for broadband service in eligible rural communities. Currently, RUS/USDA houses the only two federal assistance programs *exclusively* dedicated to financing broadband deployment: the Rural Broadband Access Loan and Loan Guarantee Program and the Community Connect Grant Program.

The 11 0th Congress considered reauthorization and modification of the program as part of the farm bill. H.R. 2419/H.R. 6124, the Food, Conservation, and Energy Act of 2008 became law on June 18, 2008 (P.L. 110-246). Title VI (Rural Development) contains authorizing language for the broadband loan program. Additionally, H.R. 2035, H.R. 2174, H.R. 2569, H.R. 2953, S. 541, S. 1032, and S. 1439 were introduced to address some of the reauthorization issues related to the RUS broadband loan program.

Meanwhile, on May 11, 2007, RUS released a Proposed Rule seeking to revise the broadband loan program rules and regulations. Some key issues pertinent to a

* Excerpted from CRS Report for Congress, Order Code RL33816, dated August 21, 2008.

consideration of the RUS broadband programs include restrictions on applicant eligibility, how "rural" is defined with respect to eligible rural communities, how to address assistance to areas with preexisting broadband service, technological neutrality, funding levels and mechanisms, and the appropriateness of federal assistance. The final rule will reflect language in the farm bill statute (P.L. 110-246). Ultimately, modification of rules, regulations, or criteria associated with the RUS broadband program will likely result in "winners and losers" in terms of which companies, communities, regions of the country, and technologies are eligible or more likely to receive broadband loans and grants.

INTRODUCTION

The Rural Utilities Service (RUS) within the U.S. Department of Agriculture (USDA) houses the only two federal assistance programs *exclusively* dedicated to financing deployment of broadband Internet access in rural America. These are: the Rural Broadband Access Loan and Loan Guarantee Program and the Community Connect Grant Program. The two programs initially appeared as pilot programs in 2001 and 2002. The broadband loan program was authorized by the 2002 farm bill (P.L. 107-17 1); this authorization expired on September 30, 2007.

The 1 10th Congress considered the RUS broadband program as part of the reauthorization of the farm bill. Given concerns over the lagging status of broadband deployment in many rural areas, Congress examined how the RUS broadband programs might be positioned to most effectively address rural broadband development. This report provides detailed background information on the RUS broadband loan and grant programs, outlines criticisms of how the RUS broadband program has been implemented thus far, and discusses issues that Congress considered during the reauthorization process.

BACKGROUND: BROADBAND AND RURAL AMERICA

The broadband loan and grant programs at RUS are intended to accelerate the deployment of broadband services in rural America. "Broadband" refers to highspeed Internet access for private homes, commercial establishments, schools, and public institutions. Currently in the United States, broadband is primarily provided via cable modem (from the local provider of cable television service) or over the telephone line (digital subscriber line or "DSL"). Other broadband technologies include fiber optic cable, fixed wireless, satellite, and broadband over power lines (BPL).

Broadband access enables a number of beneficial applications to individual users and to communities. These include e-commerce, telecommuting, voice service (voice over the Internet protocol or "VOIP"), distance learning, telemedicine, public safety, and others. It is becoming generally accepted that broadband access in a community can play an important role in economic development. A February 2006 study by the Massachusetts Institute of Technology for the Department of Commerce's Economic Development Administration marked the first attempt to measure the impact of broadband on economic growth. The study found that "between 1998 and 2002, communities in which mass-market broadband was available by December 1999 experienced more rapid growth in employment, the number of businesses overall, and businesses in IT-intensive sectors, relative to comparable communities without broadband at that time."[1]

Subsequently, a June 2007 report from the Brookings Institution found that for every one percentage point increase in broadband penetration in a state, employment is projected to increase by 0.2 to 0.3% per year. For the entire U.S. private non-farm economy, the study projected an increase of about 300,000 jobs, assuming the economy is not already at full employment.[2]

Access to affordable high-speed Internet service is viewed as particularly important for the economic development of rural areas because it enables individuals and businesses to participate fully in the online economy regardless of geographical location. For example, aside from enabling existing businesses to remain in their rural locations, broadband access could attract new business enterprises drawn by lower costs and a more desirable lifestyle. Essentially, broadband potentially allows businesses and individuals in rural America to live locally while competing globally in an online environment.

Given the large potential impact broadband may have on the economic development of rural America, concern has been raised over a "digital divide" between rural and urban or suburban areas with respect to broadband deployment. While there are many examples of rural communities with state of the art telecommunications facilities,[3] recent surveys and studies have indicated that, in general, rural areas tend to lag behind urban and suburban areas in broadband deployment. For example:

- A September 2004 Department of Commerce report, *A Nation Online: Entering the Broadband Age*, found that a lower percentage of Internet households have broadband in rural areas (24.7%) than in urban areas (40.4%), and that "while broadband usage has grown significantly in all areas since the previous survey, the rural-urban differential continues."[4] The report also found that broadband penetration rates are higher in the West and Northeast than in the South and Midwest.[5]
- A May 2006 report released by the Government Accountability Office (GAO) found that 17% of rural households subscribe to broadband, as opposed to 28% of suburban and 29% of urban households.[6] GAO also found that lower broadband subscription rates in rural areas are related to availability, not to a lesser tendency of rural households to purchase broadband service.[7]
- 2008 data from the Pew Internet & American Life Project indicate that while broadband adoption is growing in urban, suburban, and rural areas, broadband users make up larger percentages of urban and suburban users than rural users. Pew found that the percentage of all U.S. adults with broadband at home is 60% for suburban areas, 57% for urban areas, and 38% for rural areas.[8]
- According to the latest FCC data on the deployment of high-speed Internet connections (released March 2008), high-speed subscribers were reported in 99% of the most densely populated zip codes, as opposed to 91% of zip codes with the lowest population densities.[9]

The comparatively lower population density of rural areas is likely the major reason why broadband is less deployed than in more highly populated suburban and urban areas. Particularly for wireline broadband technologies — such as cable modem and DSL — the greater the geographical distances among customers, the larger the cost to serve those customers. For example, in providing telecommunications services, investment per subscriber in rural systems averages $2,921 compared to $1,920 for urban.[10] Thus, there is often less incentive for companies to invest in broadband in rural areas than, for example, in an urban area where there is more demand (more customers with perhaps higher incomes) and less cost to wire the market area.

The terrain of rural areas can also be a hindrance, in that it is more expensive to deploy broadband technologies in a mountainous or heavily forested area. An additional added cost factor for remote areas can be the expense of "backhaul" (e.g., the "middle mile") which refers to the installation of a dedicated line which transmits a signal to and from an Internet backbone which is typically located in or near an urban area.

Cable modem and DSL currently comprise over 60% of broadband deployment nationwide.[11] However, because of the challenges of deploying these technologies in low population density areas, other broadband technologies have been identified as perhaps offering potential in rural areas. These include mobile wireless (cellular), fixed wireless (WIMAX, wi-fi), satellite, and broadband over powerlines (BPL).

Pilot Broadband Loan and Grant Programs

Given the lagging deployment of broadband in rural areas, Congress and the Administration acted to initiate pilot broadband loan and grant programs within the Rural Utilities Service of the U.S. Department of Agriculture. While RUS had long maintained telecommunications loan and grant programs (Rural Telephone Loans and Loan Guarantees, Rural Telephone Bank, and more recently, the Distance Learning and Telemedicine Loans and Grants) none were exclusively dedicated to financing rural broadband deployment. Title III of the FY2001 agriculture appropriations bill (P.L. 106-387) directed USDA/RUS to conduct a "pilot program to finance broadband transmission and local dial-up Internet service in areas that meet the definition of 'rural area' used for the Distance Learning and Telemedicine Program."

Subsequently, on December 5, 2000, RUS announced the availability of $100 million in loan funding through a one-year pilot program "to finance the construction and installation of broadband telecommunications services in rural America."[12] The broadband pilot loan program was authorized under the authority of the Distance Learning and Telemedicine Program (7 U.S.C. 950aaa) , and was available to "legally organized entities" not located within the boundaries of a city or town having a population in excess of 20,000.

The FY2001 pilot broadband loan program received applications requesting a total of $350 million. RUS approved funding for 12 applications totaling $100 million. The FY2002 agriculture appropriations bill (P.L. 107-76) designated a loan level of $80 million for broadband loans, and on January 23, 2002, RUS announced that the pilot program would be extended into FY2002, with $80 million in loans made available to fund many of the applications that did not receive funding during the previous year.[13]

Meanwhile, the FY2002 agriculture appropriations bill (P.L. 107-76) allocated $20 million for a pilot broadband grant program, also authorized under the Distance Learning and Telemedicine Program. On July 8, 2002, RUS announced the availability of $20 million for a pilot grant program for the provision of broadband service in rural America. The program was specifically targeted to economically challenged rural communities with no existing broadband service. Grants were made available to entities providing "community-oriented connectivity" which the RUS defined as those entities "who will connect the critical community facilities including the local schools, libraries, hospitals, police, fire and rescue services and who will operate a community center that provides free and open access to residents."[14]

In response to the July 8, 2002, Notice of Funds Availability, RUS received more than 300 applications totaling more than $185 million in requested grant funding. RUS approved 40 grants totaling $20 million. The pilot program was extended into FY2003, as the Consolidated Appropriations Resolution of 2003 (P.L. 108-7) allocated $10 million for broadband grants. On September 24, 2003, 34 grants were awarded to eligible applicants who did not receive funding during the previous year.

Rural Broadband Access Loan and Loan Guarantee Program

Building on the pilot broadband loan program at RUS, Section 6103 of the Farm Security and Rural Investment Act of 2002 (P.L. 107-171) amended the Rural Electrification Act of 1936 to authorize a loan and loan guarantee program to provide funds for the costs of the construction, improvement, and acquisition of facilities and equipment for broadband service in eligible rural communities.[15] Section 6103 made available, from the funds of the Commodity Credit Corporation (CCC), a total of $100 million through FY2007 ($20 million for each of fiscal years 2002 through 2005, and $10 million for each of fiscal years 2006 and 2007). P.L. 107-171 also authorized any other funds appropriated for the broadband loan program.

Beginning in FY2004, Congress annually blocked mandatory funding from the CCC. Thus — starting in FY2004 — the program was funded as part of annual appropriations in the Distance Learning and Telemedicine account within the Department of Agriculture appropriations bill. Every fiscal year, Congress has approved an appropriation for the loan program which is used to subsidize a specific loan level (the total amount of lending authority). Table 1 shows — for the life of the program to date — loan subsidies, loan levels (lending authority), and actual funds announced by RUS yearly for loan applications. Announced available funding typically exceeds yearly loan levels because large balances of unobligated money have been carried over from year to year.

Table 1. Funding for the Rural Broadband Access Loan and Loan Guarantee Program

	Budget Authority (subsidy level)	Loan Level (lending authority)	Announced Available Funding for Loans and Loan Guaranteesa
FY2003	$40 millionb	$1.455 billion	$1.455 billionc
FY2004	$13.1 million	$602 million	$2.211 billiond
FY2005	$11.715 million	$550 million	$2.157 billione
FY2006	$10.75 million	$500 million	$1.085 billionf
FY2007	$10.75 million	$500 million	$0.998 billiong
FY2008	$6.45 million	$300 million	$0.7929 billionh
FY2009 (req)	$11.619 million	$300 million	Not yet available

[a] Because all available funds were not awarded, unobligated balances were carried over from year to year.

[b] Composed of $20 million from FY2002 plus $20 million for FY2003 of mandatory funding from the Commodity Credit Corporation, as directed by P.L. 107-171. In the FY2004, FY2005, and FY2006 appropriations bills, mandatory funding from the CCC was canceled.

[c] Rural Utilities Service, USDA, "Rural Broadband Access Loans and Loan Guarantees Program,"

[d] Federal Register, Vol. 68, No. 20, January 30, 2003, pp. 4753-4755.

[e] Rural Utilities Service, USDA, "Rural Broadband Access Loans and Loan Guarantees Program,"

[f] Federal Register, Vol. 69, No. 60, March 29, 2004, pp. 1623 1-16232.

[g] Rural Utilities Service, USDA, "Rural Broadband Access Loans and Loan Guarantees Program,"

[h] Federal Register, Vol. 70, No. 42, March 4, 2005, pp. 10595-10596.

[i] USDA, Rural Utilities Service, "Rural Development Utilities Programs," powerpoint presentation, April 25, 2006. Available at [http://www.usda.gov/rus/pasd/NARUC-2006April.ppt#322,1].

[g] USDA, Rural Utilities Service, "Rural Development Telecommunications Program Overview," powerpoint presentation, October 1, 2007. Available at [http://www.usda.gov/rus/telecom/broadband/workshops/J.PontiMichiganPresentationOct12007.ppt].

[k] USDA, Rural Utilities Service, "Rural Development Telecommunications Program Overview," powerpoint presentation, February 13, 2008. Available at [http://www.rurdev.usda.gov/or/DLT/Telecom-Overview-Bradley.pdf].

The Rural Broadband Access Loan and Loan Guarantee Program is codified as 7 U.S.C. 950bb. Specifically, Treasury rate loans, 4% loans, and loan guarantees are authorized for entities providing broadband service for "eligible rural communities," defined as any area of the United States that is not contained in an incorporated city or town with a population in excess of 20,000 inhabitants.[16] RUS is required to be technologically neutral in determining whether or not to make a loan, and is instructed to give priority to rural communities with no existing residential broadband service. Loans are used for financing new or improved existing broadband provider facilities. Loans cannot be used to finance installations or equipment at customers' premises.

On January 30, 2003, the RUS published in the *Federal Register* the regulation (7 C.F.R. part 1738) establishing the Rural Broadband Access Loan and Loan Guarantee Program, as authorized by P.L. 107-171.[17] According to the regulation, entities eligible to receive loans include corporations, limited liability companies, cooperative or mutual organizations, Indian tribes, and public bodies. Specifically not eligible are individuals, partnerships, and any entity serving 2% or more of the telephone subscriber lines in the United States. All applicants are required to demonstrate adequate credit support — a minimum of 20% of requested loan amount, including cash on hand equivalent to one full year of operating expense.[18]

To be eligible for 4% loans, applicants must be proposing to serve a community with no existing broadband service, a population of 2,500 or less, and a service area with population density of no more than 20 persons per square mile. Additionally, the community must be located in a county with a per capita income of less than or equal to 65% of the national per capita income.

As of February 1, 2008, the broadband loan program received 212 applications, requesting a total of $4.496 billion in loans. Of these, 79 applications were approved (totaling $1.3 billion), 19 were in review (totaling $887 million), and 113 had been returned (totaling $2.26 billion). RUS estimates that 653,000 households in 1,438 rural communities will receive broadband service as a result of approved loans. Of loans approved, 53% were made to corporations, 36% to LLCs, 7% to cooperatives, 3% to municipalities, and 1% to a tribal authority. Ten projects have completed in Kansas, Louisiana, Michigan, North Dakota, Nebraska, South Dakota, Oregon, Texas, and Washington.[19]

Subsequently, on March 25, 2008, RUS announced its largest loan ever, $267 million to Open Range Communications. The $267 million loan, accompanied by an over $100 million investment from the private sector, will enable Open Range to provide wireless broadband (Wi-Max technology) and satellite connectivity to 518 rural communities in 17 states.[20]

Applications for the Rural Broadband Access Loan and Loan Guarantee program are accepted at any time. The maximum loan amount for 4% loans is $7.5 million. There is no maximum for treasury rate loans, and the minimum level for all loans is $100 thousand. In 2003, the average loan was $11.2 million, while in 2006, the average loan was $44 million.[21] Loans are made for the term equal to the expected service life of financed facilities. Further information, including application materials and guidelines, is available at [http:// www.usda.gov/rus/telecom/broad band. htm].

COMMUNITY CONNECT BROADBAND GRANTS

The Consolidated Appropriations Act of 2004 (P.L. 108-199) appropriated $9 million "for a grant program to finance broadband transmission in rural areas eligible for Distance Learning and Telemedicine Program benefits authorized by 7 U.S.C. 950aaa." On July 28, 2004, RUS published its final rule on the broadband grant program, called the Community Connect Grant Program (7 C.F.R. part 1739, subpart A).[22] Essentially operating the same as the pilot broadband grants, the program provides grant money to applicants proposing to provide broadband on a "community-oriented connectivity" basis to currently unserved rural areas for the purpose of fostering economic growth and delivering enhanced health care, education, and public safety services.

Funding for the broadband grant program is provided through annual appropriations in the Distance Learning and Telemedicine account within the Department of Agriculture appropriations bill. Table 2 shows a history of appropriations for the Community Connect Broadband Grants (including the pilot grants of FY2002 and FY2003).

Table 2. Appropriations for the Community Connect Broadband Grants

Fiscal Year	Appropriation
FY2002	$20 million
FY2003	$10 million
FY2004	$9 million
FY2005	$9 million
FY2006	$9 million
FY2007	$9 million
FY2008	$13.5 million
FY2009 (req)	0

Source: Compiled by CRS from appropriations bills.

Eligible applicants for broadband grants include incorporated organizations, Indian tribes or tribal organizations, state or local units of government, cooperatives, private corporations, and limited liability companies organized on a for profit or not- for-profit basis. Individuals or partnerships are not eligible.

Funded projects must: serve a rural area of 20,000 population or less23 where broadband service does not exist, serve one and only one single community, deploy free basic broadband service (defined as 200 kbps in both directions) for at least two years to all community facilities, offer basic broadband to residential and business customers, and provide a community center with at least ten computer access points within the proposed service area while making broadband available for two years at no charge to users within that community center.

Since the inception of the RUS broadband grant program, $81 million in grant money has been awarded to 166 communities. Awardees must contribute a matching contribution equal to 15% of the requested grant amount.

RUS typically publishes an annual Notice of Funding Availability (NOFA) in the Federal Register, which specifies the deadline for applications, the total amount of funding available, and the maximum and minimum amount of funding available for each grant. Further information, including application materials and guidelines, is available at [http:// www.usda.gov/rus/telecom/commconnect.htm].

OTHER BROADBAND PROGRAMS

The Rural Broadband Access Loan and Loan Guarantee Program and the Community Connect Broadband Grants are currently the only federal programs *exclusively* dedicated to deploying broadband infrastructure. However, there exist other federal programs that provide financial assistance for various aspects of telecommunications development.24 Though not explicitly or exclusively devoted to broadband, many of those programs are used to help deploy broadband technologies in rural areas. For example, since 1995, the RUS Rural Telephone Loan and Loan Guarantee program — which has traditionally financed telephone voice service in rural areas under 5,000 inhabitants — has required that all telephone facilities receiving financing must be capable of providing DSL broadband service at a rate of at least 1 megabyte per second.[25] An October 2006 survey of RUS traditional telephone loan program borrowers found that 92% of those borrowers were providing broadband to all of the telephone exchanges in their service territories.[26] The RUS Distance Learning and Telemedicine grants program is used to support deployment of broadband technologies specifically for telemedicine and distance learning applications. Table 3 shows the number of customers receiving broadband due to USDA financing of telecommunications facilities.

Table 3. Number of Customers Receiving New or Improved Telecommunication Services (Broadband) Through USDA Financing of Telecommunications Facilities (millions)

FY01	FY02	FY03	FY04	FY05	FY06	FY07	FY08	FY09
0.315	0.31	0.38	0.37	0.23	0.30	0.32	0.39	0.23

Sources: U.S. Dept. of Agriculture, 2006 Performance and Accountability Report, November 2006, p. 82; U.S. Dept. of Agriculture, FY2009 Budget Summary and Performance Plan, p. 39.

Note: Customers are defined as access lines financed by the programs.

The other major vehicle for funding telecommunications development in rural areas is the Universal Service Fund (USF).[27] Subsidies provided by USF's Schools and Libraries Program and Rural Health Care Program are used for a variety of telecommunications services, including broadband access. While the USF's High Cost Program does not *explicitly* fund broadband infrastructure, subsidies are used, in many cases, to upgrade existing telephone networks. Regarding the USF High Cost Program, the Congressional Budget Office has found that "current policy implicitly provides funds for broadband in rural areas," adding that:

> Whether such upgrades are motivated by the intention to provide broadband or better conventional telephone service is not immediately clear. However, the fact that wireline carriers as a whole have been losing subscribers and long-distance revenue over the past half decade suggests that at least part of the new investment in local loops has been made with the expectation of generating revenue from broadband subscriptions.[28]

In the 110th Congress, legislation to reform universal service — which could have a significant impact on the amount of financial assistance available for broadband deployment in rural and underserved areas — has been introduced. For more information on universal service, see CRS Report RL33979, *Universal Service Fund: Background and Options for Reform*, by Angele A. Gilroy.

In addition to federal support for broadband deployment, there are programs and activities ongoing at the state and local level. Surveys, assessments, and reports from the American Electronics Association,[29] Technet,[30] the Alliance for Public Technology,[31] the California Public Utilities Commission,[32] the AEI-Brookings Joint Center,[33] and the National Conference of State Legislatures[34] have explored state and local broadband programs. A related issue is the emergence of municipal broadband networks (primarily wireless and fiber based) and the debate over whether such networks constitute unfair competition with the private sector.

CRITICISMS OF RUS BROADBAND PROGRAMS

Broadband loan and grant programs have been awarding funds to entities serving rural communities since FY2001. Since their inception, a number of criticisms of the RUS broadband loan and grant programs have emerged.

Loan Approval and Application Process

Perhaps the major criticism of the broadband loan program is that not enough loans are approved, thereby making it difficult for rural communities to take full advantage of the program. As of May 1, 2007, the broadband loan program had received 198 applications, requesting a total of $4.3 billion in loans. Of these, 69 applications were approved (totaling $1.21 billion), 21 were in review (totaling $950 million), and 108 had been returned (totaling $2.14 billion).[35] According to RUS officials, 28% of available loan money was awarded in 2004, and only 5% of available loan money was awarded in 2005.[36]

The loan application process has been criticized as being overly complex and burdensome, requiring applicants to spend months preparing costly market research and engineering assessments. Many applications are rejected because the applicant's business plan is deemed insufficient to support a commercially viable business. The biggest reason for applications being returned is insufficient credit support, whereby applicants do not have sufficient cash-on-hand (one year's worth is required in most cases). The requirement for cash-on-hand is viewed as particularly onerous for small start up companies, many of whom lack sufficient capital to qualify for the loan. Such companies, critics assert, may be those entities most in need of financial assistance.

In report language to the FY2006 Department of Agriculture Appropriations Act (P.L. 109-97), the Senate Appropriations Committee (S.Rept. 109-92) directed the RUS "to reduce the burdensome application process and make the program requirements more reasonable, particularly in regard to cash-on-hand requirements." The Committee also directed USDA to hire more full-time employees to remedy delays in application processing times.

At a May 17, 2006 hearing held by the Senate Committee on Agriculture, Nutrition, and Forestry, the Administrator of the RUS stated that RUS is working to make the program more user friendly, while at the same time protecting taxpayer investment:

> As good stewards of the taxpayers' money, we must make loans that are likely to be repaid. One of the challenges in determining whether a proposed project has a reasonable chance of success is validating the market analysis of the proposed service territory and ensuring that sufficient resources are available to cover operating expenses throughout the construction period until such a time that cash flow from operations become sufficient. The loan application process that we have developed ensures that the applicant addresses these areas and that appropriate resources are available for maintaining a viable operation.[37]

According to RUS, the loan program was initially overwhelmed by applications (particularly during a two week period in August 2003), and as the program matures, application review times have dropped.[38] On May 11, 2007, RUS released a Proposed Rule which seeks to revise regulations for the broadband loan program. In the background material accompanying the Proposed Rule, RUS stated that the average application processing time in 2006 was almost half of what it was in 2003.[39]

Eligibility Criteria

Since the inception of the broadband grant and loan programs, the criteria for applicant eligibility has been criticized both for being too broad and for being too narrow. An audit report released by USDA's Office of Inspector General (IG) found that the "programs' focus has shifted away from those rural communities that would not, without Government assistance, have access to broadband technologies."[40] Specifically the IG report found that the RUS definition of rural area has been "too broad to distinguish usefully between suburban and rural communities,"[41] with the result that, as of March 10, 2005, $103.4 million in loans and grants (nearly 12% of total funding awarded) had been awarded to 64 communities located near large cities. The report cited examples of affluent suburban subdivisions qualifying as rural areas under the program guidelines and receiving broadband loans.[42]

On the other hand, eligibility requirements have also been criticized as too narrow. For example, the limitation of assistance only to communities of 20,000 or less in population excludes small rural towns that may exceed this limit, and also excludes many municipalities seeking to deploy their own networks.[43] Similarly, per capita income requirements can preclude higher income communities with higher costs of living (e.g. rural Alaska), and the limitation of grant programs only to underserved areas excludes rural communities with existing but very limited broadband access.[44]

Loans to Communities with Existing Providers

The USDA Rural Broadband Access statute (7 U.S.C. 950bb) specifies that the program "shall give priority to eligible rural communities in which broadband service is not available to residential customers." The IG report found that RUS too often has given loans to communities with existing broadband service. The IG report found that "RUS has not ensured that communities without broadband service receive first priority for loans," and that although RUS has a system in place to prioritize loans to unserved communities, the system "lacks a cutoff date and functions as a rolling selection process — priorities are decided based on the applicants who happen to be in the pool at any given moment."[45] The result is that a significant number of communities with some level of preexisting broadband service have received loans. According to the IG report, of 11 loans awarded in 2004, 66% of the associated communities served by those loans had existing service. According to RUS, 31% of communities served by all loans (during the period 2003 through early 2005) had preexisting competitive service (not including loans used to upgrade or expand existing service).[46] In some cases, according to the IG report, "loans were issued to companies in highly competitive business environments where multiple providers competed for relatively few customers."[47] At the May 1, 2007 hearing before the House Subcommittee on Specialty Crops, Rural Development, and Foreign Agriculture, RUS Administrator James Andrews testified that of the 69 broadband loans awarded since the program's inception, 40% of the communities approved for funding were unserved at the time of loan approval, and an additional 15% had only one broadband provider.[48]

Awarding loans to entities in communities with preexisting competitive service has raised criticism from competitors who already offer broadband to those communities. According to the National Cable and Telecommunications Association (NCTA), "RUS loans are being used to unfairly subsidize second and third broadband providers in communities where private risk capital already has been invested to provide broadband service."[49] Critics argue that providing loans in areas with preexisting competitive broadband service creates an uneven playing field and discourages further private investment in rural broadband.[50] In response, RUS stated in the IG report that its policies are in accordance with the statute, and that they address "the need for competition to increase the quality of services and reduce the cost of those services to the consumer."[51] RUS argues that the presence of a competitor does not necessarily mean that an area is adequately served, and additionally, that in order for some borrowers to maintain a viable business in an unserved area, it may be necessary for that company to also be serving more densely populated rural areas where some level of competition already exists.[52]

Issues for Reauthorization

The current authorization for the Rural Broadband Access Loan and Loan Guarantee program expired on September 30, 2007. The 1 10th Congress considered reauthorization of the program as part of the farm bill. Any modification of rules, regulations, or criteria associated with the RUS broadband program will likely result in "winners and losers" in terms of which companies, communities, regions of the country, and technologies are eligible or more likely to receive broadband loans and grants.

On May 11, 2007, RUS released a Proposed Rule which seeks to revise regulations for the broadband loan program and address many of the criticisms of the program. Specifically, the Proposed Rule addresses: (1) funding in competitive markets and new eligibility requirements; (2) new equity and market survey requirements; and (3) new legal notice requirements to increase transparency.[53]

Congress considered approaches in the Proposed Rule as part of formulating the RUS broadband program reauthorization. The final rule will reflect language in the farm bill statute (P.L. 110-246). The following are some key issues pertinent to the reauthorization of the RUS broadband programs.

Restricting Applicant Eligibility

The RUS broadband program has been criticized for excluding too many applicants due to stringent financial requirements (e.g. the requirement that an applicant have a year's worth of cash-on-hand) and an application process — requiring detailed business plans and market surveys — that some view as overly expensive and burdensome to complete. During the reauthorization process, Congress considered whether the criteria for loan eligibility should be modified, and whether a more appropriate balance can be found between the need to make the program more accessible to unserved and often lower-income rural areas, and the need to protect taxpayers against bad loans.

The Proposed Rule issued by RUS on May 11, 2007, would substantially modify applicant eligibility restrictions. First, the Proposed Rule would eliminate the requirement for a market survey for applicants proposing to serve 15% or less of a community.[54] Second, the existing "credit support requirement," would be replaced by an "equity requirement." Under the Proposed Rule, the equity requirement would be reduced from 20% to 10% for applicants proposing to serve areas where at least 40% of households have no broadband access or service from only one provider. Additionally, the current requirement for a year's worth of cash-on-hand would be eliminated. Instead, once RUS has completed its review of the loan application, the applicant would be notified if additional cash requirements are needed to support the feasibility of the loan. This would be the case if RUS financial analysis indicates that cash from operations and previous cash infusions cannot sustain a positive cash position throughout the forecast period. Finally, the Proposed Rule seeks to give RUS the authority to modify or waive the provisions of the equity requirement as long as those modifications do not result in a projected negative cash position, and if those modifications are required to provide broadband in areas with no service or with only one existing broadband provider.[55]

Definition of "Rural Community"

The definition of which communities qualify as "rural" has been changed twice by statute since the broadband loan program was initiated. Under the pilot program, funds were authorized under the Distance Learning and Telemedicine Program, which defines "exceptionally rural areas" (under 5,000 inhabitants), "rural areas" (between 5,000 and 10,000) and "mid-rural areas" (between 10,000 and 20,000). RUS determined that communities of 20,000 or less would be eligible for broadband loans in cases where broadband services did not already exist.

In 2002, this definition was made narrower by the Farm Security and Rural Investment Act (P.L. 107-17 1), which designated eligible communities as any incorporated or unincorporated place with fewer than 20,000 inhabitants, and which was outside any standard metropolitan statistical area (MSA). The requirement that communities not be located within MSA's effectively prohibited suburban communities from receiving broadband loans. However, in 2004, the definition was again changed by the FY2004 Consolidated Appropriations Act (P.L. 108-199). The act broadened the definition, keeping the population limit at 20,000, but eliminating the MSA prohibition, thereby permitting rural communities near large cities to receive loans. Thus the current definition used for rural communities is the same as what was used for the broadband pilot program, except that loans can now be issued to communities with preexisting service.

The definition of what constitutes a "rural" community is always a difficult issue for Congressional policymakers in determining how to target rural communities for broadband assistance. On the one hand, the narrower the definition the greater the possibility that deserving communities may be excluded. On the other hand, the broader the definition used, the greater the possibility that communities not traditionally considered "rural" or "underserved" may be eligible for financial assistance.

The Proposed Rule released by RUS on May 11, 2007 would narrow the definition of an Eligible Rural Community. In addition to excluding cities or towns with populations over 20,000, the Proposed Rule would exclude communities located within an Urban Area, as defined by the Bureau of the Census as all territory, population, and housing units located within an urbanized area or an urban cluster. Also excluded is any area that has four or more existing broadband service providers (excluding the applicant).[56]

A related issue is the scope of coverage proposed by individual applications. While many of the loan applications propose broadband projects offering service to multiple rural communities, RUS sees a coming trend towards larger regional and national proposals, covering hundreds or even more than a thousand communities.[57] The larger the scope of coverage, the greater the complexity of the loan application and the larger the possible benefits and risks to taxpayers.

Preexisting Broadband Service

While the majority of broadband loans (and all broadband grants) are awarded to entities serving areas without preexisting broadband service, and while RUS is directed by statute to "give priority to eligible rural communities in which broadband service is not available to residential customers," a significant number of Treasury- rate loans have been awarded in areas with preexisting service. Loans to areas with competitive preexisting service — that is, areas where existing companies already provide some level of broadband — have sparked controversy because loan recipients are likely to compete with other companies already providing broadband service.

During reauthorization, Congress was asked to more sharply define whether and/or how loans should be given to companies serving rural areas with preexisting competitive service.[58] On the one hand, one could argue that the federal government should not be subsidizing competitors for broadband service, particularly in sparsely populated rural markets which may be able only to support one provider. Furthermore, keeping communities with preexisting broadband service eligible may divert assistance from unserved areas that are most in need. On other hand, many suburban and urban areas currently receive the benefits of competition between broadband providers — competition which can potentially drive down prices while improving service and performance. It is therefore appropriate, it is argued, that rural areas also receive the benefits of competition, which in some areas may not be possible without federal financial assistance. It is also argued that it may not be economically feasible for borrowers to serve sparsely populated unserved communities unless they are permitted to also serve more lucrative areas which may already have existing providers.

The Proposed Rule released by RUS on May 11, 2007, would prohibit funding for any community where there are four or more Existing Broadband Service Providers. To be recognized as an Existing Broadband Service Provider, an incumbent service provider must provide evidence and certify to RUS that 10% of the households passed by their facilities are purchasing their broadband service.[59] Applicants would be required to prepare a legal notice stating the intent to offer broadband service in a particular community. The legal notice would be published on the RUS web page. Incumbent providers would then have 30 days to submit information to RUS which would be used to determine if the incumbent is classified as an Existing Broadband Service Provider.[60]

New market entrants and start-ups, as well as incumbent applicants seeking to expand their service area, are required to enter areas where at least 40% of households either have no broadband service or access to only one broadband service provider. According to RUS:

> This requirement addresses the need to reach unserved or underserved areas while also permitting service to more lucrative areas, which may be served by up to three Existing Service Providers, in order to attract feasible loan proposals which are supportable from project revenues. Permitting service in areas with up to three Existing Service Providers addresses the need for applicants to leverage revenues from lower-cost users (typically those in more densely populated areas within a city or town) in order to provide service to rural households in higher cost areas, while excluding areas with higher levels of competition where loan feasibility is unlikely.[61]

Additionally, the Proposed Rule states that loans to incumbent service providers can be used to upgrade existing facilities without serving additional customers, as long as the upgrades enhance existing broadband service, and as long as there are no more than three other Existing Broadband Service providers in the area.[62]

Technological Neutrality

The 2002 farm bill (P.L. 107-171) directed RUS to use criteria that are "technologically neutral" in determining which projects to approve for loans. In other words, RUS is prohibited from typically valuing one broadband technology over another when assessing loan applications. As of September 2007, 35% of approved and funded projects employed fiber-to-the-home technology, 20% employed DSL, 25% fixed wireless, 19% hybrid fiber-coaxial (cable), and 1% broadband over powerlines (BPL).[63] No funding has been provided for projects utilizing satellite broadband.[64]

While decisions on funded projects are required to be technologically neutral, RUS (through the Secretary of Agriculture) does have the latitude to determine minimum required data transmission rates for broadband projects eligible for funding. According to the statute, "the Secretary shall, from time to time as advances in technology warrant, review and recommend modifications of rate-of-data transmission criteria for purposes of the identification of broadband service technologies." To date, RUS broadband loan and grant programs have required a minimum threshold of 200 kbps (kilobytes per second) in both directions (both uploading and downloading). While the 200 kbps minimum matches the standard definition of broadband that is used by the Federal Communications Commission (FCC), it is considered a low threshold that captures almost all existing broadband technology.[65]

Some have argued that the minimum threshold of 200 kbps should be raised to ensure that rural areas receive "next-generation" broadband technologies with faster data rates capable of more varied and sophisticated applications. On the other hand, significantly raising minimum data rates could exclude certain technologies — for example typical data transmission rates for fiber and some wireless technologies exceed what is offered by "current generation" technologies such as DSL and cable. Proponents of keeping the minimum threshold at a low level could argue that underserved rural areas are best served by any broadband technology that is economically feasible to deploy, regardless of whether it is "next" or "current" generation.

Funding

Under the 2002 farm bill (P.L. 107-171), broadband loan subsidies were funded at a total of $100 million through FY2007 ($20 million for each of fiscal years 2002 through 2005, and $10 million for each of fiscal years 2006 and 2007). This $100 million was to be transferred from funds of the Commodity Credit Corporation (CCC). However, beginning in FY2004, Congress has annually blocked mandatory funding from the CCC, thus ensuring that the program was funded solely through annual appropriations.

During reauthorization, the 110th Congress considered whether the mandatory CCC funding mechanism provided in the 2002 farm bill should be retained, eliminated, or modified. Also at issue was whether the current funding levels for the RUS broadband programs are optimal. Given the relatively low utilization of the broadband loan program, should funding remain at current levels or below, or alternatively, if modifications are made to ensure fuller utilization, and given the need to bridge the digital divide, should funding be increased? A related issue was whether more money should be shifted from the broadband loan program to the Community Connect broadband grant program, in order to better address the need for broadband in lower income rural communities that may not be able to meet financial criteria necessary to qualify for loans.

Appropriateness of Federal Assistance

Finally, there is the broader issue of whether government intervention in the broadband marketplace is appropriate or effective. Some argue that federal investment in broadband deployment could distort private sector investment decisions in a dynamic and rapidly evolving marketplace,[66] and question whether other strategies — such as deregulation, tax incentives, or spectrum policy — may be more effective in fostering increased broadband deployment.

On the other hand, proponents of financial assistance counter that the available data show, in general, that the private sector will invest in areas where it expects the greatest return: areas of high population density and income. Without some governmental assistance in underserved areas, they argue, it is reasonable to conclude that broadband deployment will lag behind in many rural and low income areas.[67]

ACTIVITIES IN THE 110TH CONGRESS

On January 31, 2007, Secretary of Agriculture Mike Johanns released the Administration's 2007 farm bill proposal. The Administration proposal would reauthorize the

Rural Broadband Access Loan and Loan Guarantee program and would allocate $500 million in mandatory spending to reduce the backlog in a number of Rural Development loan and grant programs, including the broadband loan program.

Meanwhile, in the May 11, 2007 issue of the *Federal Register*, RUS released its Proposed Rule seeking to revise the broadband loan program rules and regulations (7 C.F.R. Part 1738). RUS accepted public comments on the Proposed Rule through July 10, 2007. The final rule will reflect language in the farm bill statute (P.L. 110- 246).

House Farm Bill

On May 1, 2007, the House Subcommittee on Specialty Crops, Rural Development and Foreign Agriculture held a hearing on the RUS broadband programs. Testimony was heard from RUS Administrator James Andrew and public witnesses.[68] On June 6, 2007, the Subcommittee on Specialty Crops, Rural Development and Foreign Agriculture, chaired by Representative McIntyre, approved proposals for sections under its jurisdiction of H.R. 2419, the Farm Bill Extension Act of 2007. Title VII (Rural Development) of the Subcommittee markup included provisions on the Rural Broadband Access Loan and Loan Guarantee Program and the Community Connect Grant program. On July 19, 2007, the full committee approved H.R. 2419; on July 23, 2007, H.R. 2419 (H.Rept. 110-256, Part I) was reported (amended) as the Farm, Nutrition, and Bioenergy Act of 2007. H.R. 2419 was passed by the House on July 27, 2007. The following summarizes the broadband provisions (section 6023, section 6024, and section 6031), as passed by the House, which would change current law:

- Defines eligible rural community as any area of the United States that is not (1) included within the boundaries of any city, town, borough, or village, whether incorporated or unincorporated, with a population of more than 20,000; or (2) included within the urbanized area contiguous and adjacent to such a city or town.
- Prohibits loans to any community in which there are three or more incumbent service providers, unless the loan is to an incumbent service provider of the community, unless other providers in that community are notified of the application, and unless the application proposes to substantially increase the quality of broadband service and the provision of service to underserved households inside and outside the community. Incumbent service providers are defined as an entity providing service to at least 5% of the households in the service area proposed in the application. Also prohibits loans for new construction to any community in which more than 75% of households may obtain affordable broadband service, on request, from at least one incumbent service provider.
- Directs RUS to give priority, in the following order, to applications from eligible rural communities that have (i) no incumbent service provider; (ii) one incumbent service provider; or (iii) two incumbent service providers who, together, serve not more than 25% of the households in the service area proposed in the application.
- Directs RUS to ensure that the type, amount and method of security used to secure a loan or loan guarantee is commensurate to the risk involved with the loan, particularly when the loan is issued to a financially healthy, strong, and stable entity.

The Secretary shall consider reducing the loan security required in areas that do not have broadband service.

- Directs the Secretary to take steps to reduce the cost and paperwork associated with applying for a loan or loan guarantee by first-time applicants (particularly smaller and start-up Internet providers), while maintaining the ability of the Secretary to make an analysis of the risk associated with the loan.
- Allows loan recipients to decide over how many years to amortize the loan, up to 35 years, provided that the Secretary of Agriculture determines that the loans are adequately secure. In determining the loan term, the Secretary shall consider whether the recipient is or would be serving an area that is not receiving broadband services.
- States that an entity is not eligible for a loan if it serves more than 10% of the telephone subscriber lines installed in the United States. Not more than 25% of loans made in a single fiscal year may be approved for entities that serve more than 2% of the telephone subscriber lines in the United States.
- Authorizes for appropriation such sums as necessary for each of fiscal years 2008 through 2012. For each fiscal year, 10% of funding shall be reserved for entities serving eligible tribal communities. Amounts unobligated by June 30 of the fiscal year shall be made available to eligible entities in any state.
- Authorizes the Community Connect Grant Program through FY20 12 at $25 million per year. Requires a matching contribution of 15%. Priority is given to grants that will enhance community access to telemedicine and distance learning resources.
- Requires detailed program reports from RUS to the House Committee on Agriculture and the Senate Committee on Agriculture, Nutrition, and Forestry.
- Establishes a National Center for Rural Telecommunications Assessment.
- Requires USDA to submit a comprehensive rural broadband strategy to the House Committee on Agriculture and the Senate Committee on Agriculture, Nutrition, and Forestry.

Senate Farm Bill

On October 25, 2007, the Senate farm bill, the Food and Energy Security Act of 2007, was approved by the Senate Committee on Agriculture, Nutrition, and Forestry. Title VI, "Rural Development and Investment" contains provisions (primarily section 6110, "Access to broadband telecommunications services in rural areas") which address reauthorization of the Rural Broadband Access Loan and Loan Guarantee Program. The bill was reported by the Committee on November 2, 2007 (S. 2302; S.Rept. 110-220). The Food and Energy Security Act of 2007 (H.R. 2419, as amended) was passed by the Senate on December 14, 2007. The following summarizes broadband-related provisions that would change current law:

- Defines rural area as any area other than (1) a city or town that has a population of greater than 20,000; (2) an urbanized area contiguous and adjacent to a city or town with a population greater than 50,000; and (3) any collection of census blocks contiguous to each other with a housing density of more than 200 housing units per square mile.

- Prohibits loans or loan guarantees for a project in any specific area in which broadband service is offered by three or more terrestrial service providers that offer services comparable to the services proposed by the applicant. Provider is considered to offer broadband service in a rural area if it offers service at not more than average prices compared with similar services offered in the nearest urban area. An eligible entity must offer broadband service to at least 25% of households not offered broadband service by a terrestrial broadband service provider and must agree to complete buildout of broadband service proposed in application within three years after the loan is received.
- Directs Secretary to give priority to applicants that offer broadband to the greatest proportion of households that, prior to the provision of the service, had no terrestrial broadband service provider.
- Directs the Secretary to ensure that the type, amount, and method of security used to secure the loan is commensurate to the risk involved, particularly if the loan is issued to a financially-healthy, strong, and stable entity. Directs the Secretary to consider the recurring revenues of the entity at the time of application in determining an adequate level of credit support.
- Requires an eligible entity to provide a cost share not to exceed 10% of the amount of the loan or loan guarantee requested. Recurring revenues, including broadband service client revenues, may be credited toward the cost share.
- In case of substantially underserved trust areas (for example, Indian lands), where more than 20% of residents do not have broadband service, the Secretary has the authority to provide loans with interest rates as low as 2%, and may waive nonduplication restrictions, matching fund requirements, credit support requirements, or other regulations.
- Directs the Secretary to include a notice of application on its website for a period of not less than ninety days; included on the website will be the name of the applicant, a description and geographical representation of the proposed area of broadband service, a map and numerical estimate of the households that have no terrestrial broadband service in the proposed service area of the project, and any other relevant information. Prohibits the Secretary from making proprietary information available.
- Directs the Secretary to establish, by regulation, procedures to ensure prompt processing of applications. Annual reports describing processing times and explanations for extensions are required. Time limits for responding to applications are prescribed, including that the Secretary shall make a decision whether to approve the loan not later than 180 days after a complete application is submitted.
- Directs that the Secretary may require an entity projecting a subscriber base of more than 20% of the service market to submit a market survey; the Secretary may not require a market survey from an entity proposing to serve less than 20%.
- For direct loans, directs the Secretary to determine interest rates at the lower of the cost of borrowing to the Department of the Treasury for obligations of comparable maturity, or 4%.
- Directs that loan or loan guarantee may have a term not to exceed 30 years if the Secretary determines that the loan security is sufficient.
- Authorizes for appropriation $25 million for each of fiscal years 2008 through 2012.

- Establishes a National Center for Rural Telecommunications Assessment.
- Directs the Chairman of the FCC, in coordination with the Secretary, to submit an annual report to the House Committee on Energy and Commerce, House Committee on Agriculture, Senate Committee on Commerce, Science and Transportation, and Senate Committee on Agriculture, Nutrition, and Forestry describing a comprehensive rural broadband strategy.
- Directs GAO, not later than 30 months after enactment, to submit a study to the House Committee on Agriculture and the Senate Committee on Agriculture, Nutrition, and Forestry which examines how RUS makes decisions in allocating federal broadband benefits, what economic forces prompt applicants to seek federal funding rather than relying on the private market, how RUS awards impact the expansion of broadband infrastructure by the private sector; and to recommend what changes to federal policy are needed to further encourage technology expansion by private broadband service providers.
- "Connect the Nation Act." Directs the Department of Commerce to award grants encouraging state initiatives to improve broadband service. The grants would go to nonprofit organizations which would — working with state agencies — develop and implement statewide initiatives to identify, track, and map the availability and adoption of broadband services within states. Authorizes $40 million for each of fiscal years FY2008 through FY2012.

Conference Agreement and Public Law (P.L. 110-246)

On May 14, 2008, the farm bill conference report (H.Rept. 110-627) was approved by the House. The conference report was approved by the Senate on May 15, 2008. With respect to the broadband loan program, the conference agreement adopted much of the Senate amendment with modifications, while the definition of incumbent service provider was retained from the House bill. The conference agreement retained the provision establishing a National Center for Rural Telecommunications Assessment and the provision requiring the FCC and RUS to formulate a comprehensive rural broadband strategy. The conference agreement did *not* retain provisions related to establishing a broadband mapping and state grant program at the Department of Commerce ("Connect the Nation Act"), the GAO assessment of the RUS broadband programs, and authorization language for the Community Connect Broadband Grant program. The Food, Conservation, and Energy Act of 2008 became law on June 18, 2008 (P.L. 110-246). The following summarizes broadband-related provisions that changes previous law:

- Defines rural area as any area other than (1) a city or town that has a population of greater than 20,000 and (2) an urbanized area contiguous and adjacent to a city or town with a population greater than 50,000. The Secretary may, by regulation only, consider not to be rural an area that consists of any collection of census blocks contiguous to each other with a housing density of more than 200 housing units per square mile and that is contiguous with or adjacent to an existing boundary of a rural area.

- Prohibits the Secretary from making a loan in any area where there are more than 3 incumbent service providers unless the loan meets all of the following requirements: (1) the loan is to an incumbent service provider that is upgrading service in that provider's existing territory; (2) the loan proposes to serve an area where not less than 25% of the households are offered service by not more than 1 provider; and (3) the applicant is not eligible for funding under another provision of the Rural Electrification Act. Incumbent service provider is defined as an entity providing broadband service to not less than 5% of the households in the service territory proposed in the application. Also prohibits the Secretary from making a loan in any area where not less than 25% of the households are offered broadband service by not more than 1 provider unless a prior loan has been made in the same area.
- Provides that the highest priority is to be given to applicants that offer to provide broadband service to the greatest proportion of households currently without broadband service. Eligible entities are required to submit a proposal to the Secretary that meets the requirements for a project to offer to provide service to a rural area and agree to complete buildout of the broadband service within three years.
- Prohibits any eligible entity that provides telecommunications or broadband service to at least 20% of the households in the United States from receiving an amount of funds under this section for a fiscal year in excess of 15% of the funds authorized and appropriated for the broadband loan program.
- Allows the Secretary to require an entity to provide a cost-share in an amount not to exceed 10% of the amount of the loan or loan guarantee.
- In case of substantially underserved trust areas (for example, Indian lands), where the Secretary determines a high need exists for the benefits of the program, the Secretary has the authority to provide loans with interest rates as low as 2% and may waive nonduplication restrictions, matching fund requirements, credit support requirements, or other regulations.
- Allows the Secretary to require an entity that proposes to have a subscriber projection of more than 20% of the broadband service market in a rural area to submit a market survey. However, the Secretary is prohibited from requiring a market survey from an entity that projects to have less than 20% of the broadband market.
- Requires public notice of each application submitted, including the identity of the applicant, the proposed area to be served, and the estimated number of households in the application without terrestrial-based broadband. Authorizes the Secretary to take steps to reduce the costs and paperwork associated with applying for a loan or loan guarantee under this section by first-time applicants, particularly those who are smaller and start-up Internet providers.
- Allows the Secretary to establish a pre-application process under which a prospective applicant may seek a determination of area eligibility. Provides that an application, or a petition for reconsideration of a decision on such an application, that was pending on the date 45 days before enactment of this act and that remains pending on the date of enactment of this act is to be considered under eligibility and feasibility criteria in effect on the original date of submission of the application.

- Retains the current law rate of interest for direct loans — which is the rate equivalent to the cost of borrowing to the Department of Treasury for obligations of comparable maturity or 4%.
- Directs that loan or loan guarantee may have a term not to exceed 35 years if the Secretary determines that the loan security is sufficient.
- Directs the Secretary to consider existing recurring revenues at the time of application in determining an adequate level of credit support. Requires the Secretary to ensure that the type, amount, and method of security used to secure a loan or loan guarantee is commensurate to the risk involved with the loan or loan guarantee, particularly when the loan or loan guarantee is issued to a financially healthy, strong, and stable entity. The Secretary is also required, in determining the amount and method of security, to consider reducing the security in areas that do not have broadband service.
- Requires that the Secretary annually report to Congress on the rural broadband loan and loan guarantee program. The annual report is to include information pertaining to the loans made, communities served and proposed to be served, speed of broadband service offered, types of services offered by the applicants and recipients, length of time to approve applications submitted, and outreach efforts undertaken by USDA.
- Authorizes the Rural Broadband Access Loan and Loan Guarantee program at $25,000,000 to be appropriated for each of fiscal years 2008 through 2012.
- Provides for a National Center for Rural Telecommunications Assessment. The Center is to assess the effectiveness of broadband loan programs, work with existing rural development centers to identify appropriate policy initiatives, and provide an annual report that describes the activities of the Center, the results of research carried out by the Center, and any additional information that the Secretary may request. An appropriation of $ 1,000,000 is authorized for each of the fiscal years 2008 through 2012.
- Directs the Chairman of the FCC, in coordination with the Secretary, to submit to Congress a report describing a comprehensive rural broadband strategy. Requires the report to be updated during the third year after enactment.

Appropriations

FY2008. The President's FY2008 budget proposal requested a $6.45 million (subsidy) to support a loan level of $300 million. Noting that this is a $200 million reduction from the FY2007 level, the budget documents stated that the "funding is sufficient to meet expected demand," and that

> Regulations are being changed to correct certain weaknesses that have become apparent since the program was established a few years ago. The new regulations will ensure that program funds are focused on rural areas that are lacking existing providers, and that applicants meet high enough standards to ensure long term success.[69]

The FY2008 budget proposal requested no funding for the Community Connect Broadband Grant program.

On July 19, 2007, the House Appropriations Committee approved the FY2008 Agriculture, Rural Development, Food and Drug Administration, and Related Agencies appropriations bill (H.R. 3161; H.Rept. 110-258). The Committee approved $6.45 million to support a loan level of $300 million for the broadband loan program, and $17.82 million for broadband community connect grants (twice the FY2007 level). In report language, the Committee expressed concern over broadband loans to areas with existing providers, and directed the USDA Office of the Inspector General to conduct a comprehensive follow-up study reexamining the RUS broadband loan program. Specifically, the report is directed to detail: how many unserved households were included in approved RUS Broadband Loan Program applications; how many applications were granted to applicants proposing to serve areas where one or more private broadband providers already offered service; how many approved loans (and their total amount) have defaulted since the program's inception; and how many applicants who have been approved for loans have subsequently withdrawn from the program due to the eventually discovered infeasibility of the approved project.

In report language, the House Appropriations Committee also expressed concern over the administration of the broadband loan program, and noted USDA's failure to obligate available resources to fund broadband projects. According to the Committee, USDA will carry over $ 10.643 million from FY2007 to FY2008, which will support an additional program (loan) level of $495 million. Regarding the proposed RUS broadband loan program rule, the Committee "expects the Department to prioritize deployment of broadband service to households with no or limited broadband access." Finally, the Committee recommended $250,000 to the USDA Economic Research Service to research deployment of broadband service to households with no or limited access to broadband, and to study the economic impact of not having broadband on rural communities and their growth, community facilities, access to healthcare, and well being.

Also on July 19, 2007, the Senate Appropriations Committee approved its version of the FY2008 agriculture appropriations bill (S. 1859; S.Rept. 110-134). The Committee approved $ 10.643 million to support a loan level of $495 million for the broadband loan program, and $8.9 million for broadband grants.

On December 26, 2007, the President signed the Consolidated Appropriations Act, 2008 (P.L. 110-161). The bill provided $6.45 million to support a loan level of $300 million for the broadband loan program, and $13.5 million for broadband community connect grants. The Joint Explanatory Statement accompanying P.L. 110-161 directed USDA to evaluate and report on the potential of a combination loan/grant broadband program to expand the reach and more effectively utilize broadband resources.

FY2009

The President's FY2009 budget proposal requested a $11.61 9 million (subsidy) to support a loan level of $297.923 million. The budget justification noted that available funding in recent years has significantly outpaced demand and that $495 million was carried forward to FY2008, in addition to the FY2008 funding.

The Administration proposed a rescission of the FY2008 subsidy of $6.45 million, an unobligated balance that would otherwise be carried forward into FY2009. According to the budget justification, the program "routinely carries over the entire year's appropriations, so the funds are not necessary to support the program's demand."

As in past budget requests, the Administration proposed no funding for the Community Connect Broadband Grants (down from $13.5 million in FY2008). According to the budget justification, "building broadband infrastructure in rural America is expensive and requires extensive amounts of capital. A loan program is more suited to supporting this kind of activity."

On July 21, 2008, the Senate Appropriations Committee reported the FY2009 agriculture appropriations bill (S. 3289; S.Rept. 110-426). The Committee approved $1 1.618 million to support a loan level of $297.923 million for the broadband loan program, the same level as requested by the Administration. Unlike the Administration, the Committee provides $ 13.406 million for broadband grants.

LEGISLATION IN THE 110TH CONGRESS

P.L. 110-234 (H.R. 2419)

Farm, Nutrition, and Bioenergy Act of 2007. Reauthorizes broadband program at the Rural Utilities Service through FY20 12. Introduced May 22, 2007; referred to Committee on Agriculture, and in addition to Committee on Foreign Affairs. Reported by House Committee on Agriculture (H.Rept. 110-256) on July 23, 2007. Passed by House, July 27, 2007. Passed by Senate with an amendment, December 14, 2007. Conference report (H.Rept. 110-627) approved by the House May 14, 2008, and by the Senate May 15, 2008. Vetoed by President, May 21, 2008. House and Senate overrode veto, May 21 and May 22, 2008. Became Public Law May 22, 2008.

P.L. 110-246 (H.R. 6124)

Food, Conservation, and Energy Act of 2008. Reauthorizes broadband program at the Rural Utilities Service through FY20 12. Passed House May 22, 2008; passed Senate June 5, 2008. Vetoed by President June 18, 2008. House and Senate overrode veto June 18, 2008. Became Public Law June 18, 2008.

H.R. 2035 (Herseth Sandlin)

Rural Broadband Improvement Act. Narrows the definition of "eligible rural community" in order that communities in suburban areas and near urban areas would no longer be eligible for broadband loans. Specifically, an eligible rural community would be defined as any area that is not included within a community of more than 15,000 inhabitants, that is not included within the boundaries of an urbanized area or urban cluster, and that is not located within 10 miles of any such city, town, village, or borough, or any urbanized area of urban cluster. Additionally, seeks to limit loans awarded to applicants proposing to serve areas that already have a broadband provider. Borrowers serving eligible communities that are at least 50% already served would only be eligible for an RUS broadband loan equivalent to the portion of households that remain unserved. Introduced April 25, 2007; referred to the Committee on Agriculture and the Committee on Energy and Commerce.

H.R. 2174 (Salazar)

Rural Broadband Initiative Act of 2007. Establishes an Office of Rural Broadband Initiatives within the Department of Agriculture which will administer all rural broadband grant and loan programs previously administered by the Rural Utilities Service. Also establishes a National Rural Broadband Innovation Fund which would fund experimental and pilot rural broadband projects and applications. Introduced May 3, 2007; referred to Committee on Agriculture and to Committee on Energy and Commerce.

H.R. 2569 (Graves)

Rural Broadband Deployment Act. Codifies certain changes proposed by USDA to the rules governing eligibility for the rural broadband access program. Specifically, would relax market survey requirements and eliminate the credit support requirement, including the cash-on-hand requirement. Introduced June 5, 2007; referred to Committee on Agriculture, and in addition to the Committee on Energy and Commerce.

H.R. 2720 (Kind)

FARM 21 Act of 2007. Amends the Farm Security and Rural Investment Act of 2002 to direct that the Secretary of USDA shall make available funds of the Commodity Credit Corporation to the rural broadband loan program as follows: $10 million for each of fiscal years 2008 through 2012. Also specifies criteria to be applied by USDA in considering applications for all rural development projects. Introduced June 14, 2007; referred to Committee on Agriculture, and in addition to the Committees on Education and Labor, Foreign Affairs, and Ways and Means.

H.R. 2764 (Lowey)

Consolidated Appropriations Act, 2008. Provides $6.45 million to support a loan level of $300 million for the broadband loan program, and $13.5 million for broadband community connect grants. Signed by President, December 26, 2007.

H.R. 2953 (Space)

Rural Broadband Access Enhancement Act. Seeks to redefine "eligible rural community, streamline application process and lowers equity requirements, restricts loans to communities with existing broadband providers, eliminates limitation on eligibility based on number of subscriber lines, sets 35 year maximum on term of loan repayment, and directs USDA/RUS to meet specific reporting requirements. Introduced July 10, 2007; referred to Committee on Agriculture and Committee on Energy and Commerce.

H.R. 3161 (DeLauro)

Agriculture, Rural Development, Food and Drug Administration, and Related Agencies Appropriations, 2008. Provides $6.45 million to support a loan level of $300 million for the broadband loan program, and $17.82 million for broadband community connect grants. Introduced July 24, 2007; referred to Committee on Appropriations. Reported by Committee on Appropriations, July 24, 2007 (H.Rept. 110-258; placed on Union Calendar.

H.R. 5682 (Allen)

Rural America Communication Expansion for the Future Act of 2008. Reforms and reauthorizes through FY2013 the Rural Broadband Access Loan and Loan Guarantee Program and the Community Connect Grant Program. Introduced April 2, 2008; referred to Committee on Energy and Commerce and in addition to the Committees on Ways and Means and Agriculture.

S. 541 (Feingold)

Rural Opportunities Act of 2007. Directs the FCC to collect more detailed broadband deployment data and to periodically revise its definition of broadband above 200 kbps. Directs the Secretary of Agriculture to report on the adoption or planned adoption of the recommendations contained in the September 2005 audit report by the Inspector General of the United States Department of Agriculture. Introduced February 8, 2007; referred to Committee on Agriculture, Nutrition and Forestry.

S. 1032 (Clinton)

Rural Broadband Initiative Act of 2007. Establishes an Office of Rural Broadband Initiatives within the Department of Agriculture which will administer all rural broadband grant and loan programs previously administered by the Rural Utilities Service. Also establishes a National Rural Broadband Innovation Fund which would fund experimental and pilot rural broadband projects .and applications. Introduced March 29, 2007; referred to Committee on Agriculture, Nutrition, and Forestry.

S. 1439 (Roberts)

Rural Broadband Improvement Act of 2007 (Section 1). Section 2 reauthorizes the RUS broadband program through 2012, and makes available from funds of the Commodity Credit Corporation $20 million for fiscal years 2008 through 2010, and $10 million for fiscal years 2011 and 2012. Section 3 clarifies an "eligible rural area" as any area other than a city or town with a population greater than 20,000, the urbanized area contiguous and adjacent to such a city or town, and an area determined by the Secretary of Agriculture to be predominantly urban in character. Additionally, loans are not allowed unless the Secretary determines that no duplication of terrestrial lines, facilities, or systems would result. However, the Secretary can waive the prohibition on duplication if at least 75% of end users served by the proposed project are without broadband access, and if it is determined that such duplication is necessary in order to furnish or improve broadband service in an eligible rural area. Loan applicants must submit to RUS broadband deployment data for each existing incumbent broadband service provider in the proposed project service area. A 30-day public comment period will be initiated for each proposed loan.

Section 4 accelerates the loan application process by requiring a 180-day processing period while simplifying and/or eliminating certain loan application requirements, and authorizes RUS to hire additional staff, if necessary, to process loan applications. Section 5 eliminates the current statute's prohibition on awarding loans to any entity serving 2% or more of the telephone subscriber lines in the United States. Section 6 establishes a $20 million per year grant program at RUS which would be awarded to state-selected nonprofit organizations in order to help states develop and implement initiatives to identify and track the availability and adoption of broadband services within rural areas. Section 7 reduces the current 20% credit support requirement to a 10% equity requirement for broadband loan applicants. This requirement may be modified or waived if the applicant meets certain standards of financial feasibility. Section 8 sets a maximum term of the loan at 35 years, and Section 9 clarifies the technological neutrality standard as including DSL, fiber-optic cable, coaxial cable, wireless technology, and satellite. Introduced May 21, 2007; referred to Committee on Agriculture, Nutrition, and Forestry.

S. 1859 (Kohl)

Agriculture, Rural Development, Food and Drug Administration, and Related Agencies Appropriations, 2008. Provides $ 10.643 million to support a loan level of $495 million for the broadband loan program, and $8.9 million for broadband grants. Introduced July 24, 2007; referred to Committee on Appropriations. Reported to Senate (S.Rept. 110-134) July 24, 2007; placed on Senate Legislative Calendar.

S. 2302 (Harkin)

Food and Energy Security Act of 2007. Reauthorizes broadband program at the Rural Utilities Service through FY20 12. Introduced November 2, 2007. Senate Committee on

Agriculture, Nutrition, and Forestry reported measure to Senate (S.Rept. 110-220) November 2, 2007; placed on Senate Legislative Calendar.

Endnotes

[1] Gillett, Sharon E., Massachusetts Institute of Technology, *Measuring Broadband's Economic Impact*, report prepared for the Economic Development Administration, U.S. Department of Commerce, February 28, 2006, p. 4. Available at [http://www.eda.gov/ Image Cache/EDAPublic/documents/pdfdocs2006/mitcmubbimpactreport_2epdf/v1/mitcmubbi mpactreport.pdf].

[2] Crandall, Robert, William Lehr, and Robert Litan, *The Effects of Broadband Deployment on Output and Employment: A Cross-sectional Analysis of U.S. Data*, June 2007, 20 pp. Available at [http://www3 .brookings.edu/views/papers/crandall/200706litan.pdf].

[3] See for example: National Exchange Carrier Association (NECA), *Trends 2006: Making Progress With Broadband*, 2006, 26 p. Available at [http://www.neca.org/media/trends_ brochure_website.pdf].

[4] U.S. Department of Commerce, Economics and Statistics Administration, National Telecommunications and Information Administration, *A Nation Online: Entering the Broadband Age*, September 2004, pp. 12-13.

[5] Ibid., p. 12.

[6] U.S. Government Accountability Office, *Broadband Deployment is Extensive throughout the United States, but It Is Difficult to Assess the Extent of Deployment Gaps in Rural Areas*, GAO-06-426, May 2006, p. 12. Available at [http://www.gao.gov/new.items/ d06426.pdf].

[7] Ibid., p. 5.

[8] Horrigan, John B., Pew Internet & American Life Project, *Home Broadband Adoption 2008,* July 2008, p. 3. Available at [http://www.pewinternet.org/pdfs/PIP_Broadband_2008.pdf]

[9] FCC, *High-Speed Services for Internet Access: Status as of June 30, 2007,* p.4.

[10] Office of Management and Budget, Program Assessment Rating Tool (PART), Department of Agriculture PART Assessments, assessment year 2005, p. 262, available at [http://www.whitehouse.gov/omb/budget/fy2005/pma/agriculture.pdf].

[11] *High-Speed Services for Internet Access: Status as of June 30, 2007,* Chart 2.

[12] Rural Utilities Service, USDA, "Construction and Installation of Broadband Telecommunications Services in Rural America; Availability of Loan Funds," *Federal Register*, Vol. 65, No. 234, December 5, 2000, p. 75920.

[13] Rural Utilities Service, USDA, "Broadband Pilot Loan Program," *Federal Register*, Vol. 67, No. 15, January 23, 2002, p. 3140.

[14] Rural Utilities Service, USDA, "Broadband Pilot Grant Program," *Federal Register*, Vol. 67, No. 130, July 8, 2002, p. 45080.

[15] Title VI of the Rural Electrification Act of 1936 (7 U.S.C. 950bb).

[16] Section 772 of the FY2004 Consolidated Appropriations Act (P.L. 108-199) changed the definition of an "eligible rural community" to be defined as "any area of the United

States that is not contained in an incorporated city or town with a population in excess of 20,000 inhabitants." Accordingly, the March 29, 2004 Notice of Funds Availability for the Rural Broadband Access Loans and Loan Guarantee Program defined "Eligible Rural Community" as follows:

> The definition of eligible rural community in Section 601 (b)(2) of the Rural Electrification Act (7 U.S.C. 950bb)(b)(2), qualifying for financial assistance under the Rural Broadband Access Loan and Loan Guaranty Program, has been amended by provisions in the Consolidated Appropriations Act, 2004, to mean any area of the United States that is not contained in an incorporated city or town with a population in excess of 20,000 inhabitants. Therefore, an applicant no longer must demonstrate that it is not located in an area designated as a standard metropolitan statistical area. This change supersedes and nullifies contrary provisions in regulations implementing the broadband program found at 7 CFR part 1738.

[17] Rural Utilities Service, USDA, "Rural Broadband Access Loans and Loan Guarantees," *Federal Register*, Vol. 68, No. 20, January 30, 2003, pp. 4684-4692.

[18] The cash-on-hand requirement is waived for companies with two previous years of positive cash flow.

[19] USDA, Rural Utilities Service, "Rural Development Telecommunications Program Overview," powerpoint presentation, February 13, 2008. Available at [http:// www.rurdev.usda.gov/or/DLT/Telecom-Overview-Bradley.pdf].

[20] USDA, *News Release*, "USDA Announces $267 Million Rural Broadband Loan," March 25, 2008, available at [http://www.rurdev.usda.gov/rd/newsroom/2008/RD_Broadband_ Loans_3-25-2008.pdf].

[21] Rural Utilities Service, Department of Agriculture, "Rural Broadband Access Loans and Loan Guarantees," Proposed Rule, *Federal Register*, Vol. 72, No. 91, May 11, 2007, p. 26744.

[22] Rural Utilities Service, USDA, "Broadband Grant Program," 7 C.F.R. part 1739, *Federal Register*, Vol. 69, No. 144, July 28, 2004, pp. 44896-44903.

[23] A rural area is defined as "any area of the United States not included within the boundaries of any incorporated or unincorporated city, village, or borough having a population in excess of 20,000 inhabitants." (7 C.F.R. 1739.3)

[24] See CRS Report RL307 19, *Broadband Internet Access and the Digital Divide: Federal Assistance Programs, by* Lennard G. Kruger and Angele A. Gilroy.

[25] In the Rural Electrification Loan Restructuring Act (P.L. 103-129, the 1993 farm bill), Congress amended the Rural Electrification Act to require that facilities financed under this program be capable of providing broadband service at the rate of 1 megabyte per second (7 U.S.C. 935(d)(3)(B)(iv)(I)(cc).

[26] USDA, Rural Utilities Service, Rural Development Telecommunications home page, [http://www.usda.gov/rus/telecom/].

[27] For more information on the Universal Service Fund, see CRS Report RL307 19, *Broadband Internet Access and the Digital Divide: Federal Assistance Programs, by* Lennard G. Kruger and Angele A. Gilroy.

[28] Congressional Budget Office, *Factors That May Increase Future Spending from the Universal Service Fund*, CBO Paper, June 2006, p. 25. Available at [http://www. cbo.gov/ftpdocs/72xx/doc729 1/06-1 6-UniversalService.pdf].

[29] American Electronics Association, *Broadband in the States 2003: A State-by-State Overview of Broadband Deployment*, May 22, 2003. Available at [http://www.aeanet.org/publications/idet_broadbandstates03 .asp].

[30] TechNet, *The State Broadband Index: An Assessment of State Policies Impacting Broadband Deployment and Demand*, July 17, 2003, 48 p. Available at [http://www.analysys.com/pdfs/technetbroadband.pdf].

[31] Alliance for Public Technology, *A Nation of Laboratories: Broadband Policy Experiments in the States*, March 5, 2004, 48 p. Available at [http://www.apt.org/publications/reports-studies/broadbandreport_final.pdf].

[32] California Broadband Task Force, *The State of Connectivity: Building Innovation Through Broadband*, Final Report of the California Broadband Task Force, January 2008, 83 p. Available at [http://www.calink.ca.gov/pdf/CBTF_FINAL_Report.pdf].

[33] Wallsten, Scott, AEI-Brookings Joint Center for Regulatory Studies, *Broadband Penetration: An Empirical Analysis of State and Federal Policies*, Working Paper 05-12, June 2005, 29 p. Available at [http://aei-brookings.org/admin/authorpdfs/page.php?id=116 1].

[34] For a summary of selected state broadband bills, see [http://www.ncsl.org/programs/telecom/broadband0906.htm]

[35] Testimony of James Andrew, Administrator, Rural Utilities Service, U.S. Department of Agriculture, before the Subcommittee on Specialty Crops, Rural Development, and Foreign Agriculture, House Committee on Agriculture, May 1, 2007. Available at [http://agriculture.house.gov/testimony/1 1 0/h7050 1b/Andrew.doc]. A listing of approved and pending broadband loan applications is available at [http://www.usda.gov/rus/ telecom/broadband.htm].

[36] GAO, *Broadband Deployment is Extensive throughout the United States, but It Is Difficult to Assess the Extent of Deployment Gaps in Rural Areas*, p. 33.

[37] Testimony of Jim Andrew, Administrator, Rural Utilities Service, U.S. Department of Agriculture, "Broadband Program Administered by USDA's Rural Utilities Service," full committee hearing before the Senate Committee on Agriculture, Nutrition, and Forestry, 109th Congress, May 17, 2006.

[38] Rural Utilities Service, private communication, January 18, 2007.

[39] Rural Utilities Service, Department of Agriculture, "Rural Broadband Access Loans and Loan Guarantees," Proposed Rule, *Federal Register*, Vol. 72, No. 91, May 11, 2007, p. 26744.

[40] U.S. Department of Agriculture, Office of Inspector General, Southwest Region, *Audit Report: Rural Utilities Service Broadband Grant and Loan Programs*, Audit Report 09601-4-Te, September 2005, p. I. Available at [http://www.usda.gov/oig/webdocs/09601-04- TE.pdf].

[41] Ibid., p. 6.

[42] Ibid., p. 8.

[43] Martinez, Michael, "Broadband: Loan Fund's Strict Rules Foil Small Municipalities," *National Journal's Technology Daily*, August 23, 2005.

[44] GAO, *Broadband Deployment is Extensive throughout the United States, but It Is Difficult to Assess the Extent of Deployment Gaps in Rural Areas*, p. 33-34.

[45] Ibid., p. 13.

[46] Ibid., p. 14.

[47] Ibid., p. 15

[48] Testimony of James Andrew, Administrator, Rural Utilities Service, U.S. Department of Agriculture, before the Subcommittee on Specialty Crops, Rural Development, and Foreign Agriculture, House Committee on Agriculture, May 1, 2007.

[49] Letter from Kyle McSlarrow, President and CEO, National Cable & Telecommunications Association to the Honorable Mike Johanns, Secretary of the U.S. Department of Agriculture, May 16, 2006.

[50] Testimony of Tom Simmons, Vice President for Public Policy, Midcontinent Communications, before Senate Committee on Agriculture, Nutrition, and Forestry, May 17, 2006.

[51] *Audit Report: Rural Utilities Service Broadband Grant and Loan Programs*, p. 17.

[52] Rural Utilities Service, private communication, January 18, 2007.

[53] Rural Utilities Service, Department of Agriculture, "Rural Broadband Access Loans and Loan Guarantees," Proposed Rule, *Federal Register*, Vol. 72, No. 91, May 11, 2007, p. 26744.

[54] Ibid., p. 26750.

[55] Ibid., p. 26754.

[56] Ibid., p. 26751.

[57] Rural Utilities Service, private communication, January 18, 2007.

[58] The statute (7 U.S.C. 950bb) allows States and local governments to be eligible for loans only if "no other eligible entity is already offering, or has committed to offer, broadband services to the eligible rural community."

[59] Rural Utilities Service, Department of Agriculture, "Rural Broadband Access Loans and Loan Guarantees," Proposed Rule, *Federal Register*, Vol. 72, No. 91, May 11, 2007, p. 26749.

[60] Ibid., p. 26755.

[61] Ibid., p. 26749.

[62] Ibid.

[63] USDA, Rural Utilities Service, "Rural Development Telecommunications Program Overview," powerpoint presentation, October 1, 2007. Available at [http://www. usda.gov/rus/telecom/broadband/workshops/J.PontiMichiganPresentationOct1 2007 .ppt].

[64] According to the GAO, satellite companies state that RUS's broadband loan program requirements "are not readily compatible with their business model or technology," and that "because the agency requires collateral for loans, the program is more suited for situations where the providers, rather than individual consumers, own the equipment being purchased through the loan. Yet, when consumers purchase satellite broadband, it is common for them to purchase the equipment needed to receive the satellite signal, such as the reception dish." Satellite companies argue that in some rural areas, satellite broadband might be the most feasible and cost-effective solution. See GAO, *Broadband Deployment is Extensive throughout the United States, but It Is Difficult to Assess the Extent of Deployment Gaps in Rural Areas*, pp. 34-35.

[65] Critics of the FCC's broadband definition of 200 kbps have pointed to higher download and upload speeds typically offered in other countries. See Turner, Derek S., Free Press, *Broadband Reality Check II: The Truth Behind America's Digital Divide*, August 2006, pp 5-9. Available at [http://www.freepress.net/docs/bbrc2-final.pdf]. For further discussion of international broadband speeds and prices, including the differences

between advertised and actual speeds, see Kende, Michael, Analysis Consulting Limited, *Survey of International Broadband Offerings*, October 4, 2006, 12 p. Available at [http://www.analysys.com/pdfs/BroadbandPerformanceSurvey.pdf].

[66] See Leighton, Wayne A., *Broadband Deployment and the Digital Divide: A Primer*, a Cato Institute Policy Analysis, No. 410, August 7, 2001, 34 pp. Available at [http://www.cato.org/pubs/pas/pa410.pdf].

[67] See for example: Cooper, Mark, Consumer Federation of America and Consumers Union, *Expanding the Digital Divide & Falling Behind on Broadband*, October 2004, 33 pages. Available at [http://www.consumersunion.org/pub/ddnewbook.pdf].

[68] Testimony is available at [http://agriculture.house.gov/hearings/statements.html].

[69] U.S. Dept. of Agriculture, *FY2008 Budget Summary and Performance Plan*, p. 44.

In: Maximizing Broadband Services to Rural Communities ISBN 978-1-60692-846-2
Editors: V.B. Haynesworth, A.R. Harrison, pp.101-150

Chapter 5

TELECOMMUNICATIONS: BROADBAND DEPLOYMENT IS EXTENSIVE THROUGHOUT THE UNITED STATES, BUT IT IS DIFFICULT TO ASSESS THE EXTENT OF DEPLOYMENT GAPS IN RURAL AREAS*

United States Government Accountability Office

WHAT GAO FOUND

About 30 million American households have adopted broadband service, but the Federal Communications Commission's (FCC) data indicating the availability of broadband networks has some weaknesses. FCC conducts an extensive data collection effort using its Form 477 to assess the status of advanced telecommunications service in the United States. For its zip-code level data, FCC collects data based on where *subscribers* are served, not where providers have deployed broadband infrastructure. Although it is clear that the deployment of broadband networks is extensive, the data may not provide a highly accurate depiction of local deployment of broadband infrastructures for residential service, especially in rural areas.

A variety of market and technical factors, government efforts, and access to resources at the local level have influenced the deployment of broadband infrastructure. Areas with low population density and rugged terrain, as well as areas removed from cities, are generally more costly to serve than are densely populated areas and areas with flat terrain. As such, deployment tends to be less developed in more rural parts of the country. Technical factors can also affect deployment. GAO also found that a variety of federal and state efforts, and access to resources at the local level, have influenced the deployment of broadband infrastructure.

* Excerpted from GAO report GAO-06-425, dated May 2006.

A variety of characteristics related to households and services influence whether consumers adopt broadband service. GAO found that consumers with high incomes and college degrees are significantly more likely to adopt broadband. The price of broadband service remains a barrier to adoption for some consumers, although prices have been declining recently. The availability of applications and services that function much more effectively with broadband, such as computer gaming and file sharing, also influences whether consumers purchase broadband service.

Stakeholders identified several options to address the lack of broadband in certain areas. Although the deployment of broadband is widespread, some areas are not served, and it can be costly to serve highly rural areas. Targeted assistance might help facilitate broadband deployment in these areas. GAO found that stakeholders have some concerns about the structure of the Rural Utilities Service's broadband loan program. GAO was also told that modifications to spectrum management might address the lack of broadband infrastructure in rural areas. Also, because the cost of building land-based infrastructure is so high in some rural areas, satellite industry stakeholders noted that satellite broadband technology may be the best for addressing a lack of broadband in those regions. While several options such as these were suggested to GAO, each has some challenges to implementation. Also, a key difficulty for analyzing and targeting federal aid for broadband is a lack of reliable data on the deployment of networks.

WHY GAO DID THIS STUDY

Both Congress and the President have indicated that access to broadband for all Americans is critically important. Broadband is seen as a critical economic engine, a vehicle for enhanced learning and medicine, and a central component of 21st century news and entertainment. As part of our response to a mandate included in the Internet Tax Nondiscrimination Act of 2004, this report examines the factors that affect the deployment and the adoption of broadband services. In particular, this report provides information on (1) the current status of broadband deployment and adoption; (2) the factors that influence the deployment of broadband networks; (3) the factors that influence the adoption, or

purchase, of broadband service by households; and (4) the options that have been suggested to spur greater broadband deployment and adoption.

WHAT GAO RECOMMENDS

GAO recommends that FCC develop information regarding the cost and burden that would be associated with various options for improving the information available on broadband deployment and report this information to the relevant Senate and House committees to help them determine what actions, if any, are necessary. FCC provided technical comments on this report, but did not comment on this recommendation.

www.gao.gov/cgi-bin/getrpt?GAO-06-426.

ABBREVIATIONS

3G	third generation
ADSL	asymmetric digital subscriber line
BPL	broadband over power lines
CTIA	Cellular Telecommunications and Internet Association
DBS	direct broadcast satellite
DSL	digital subscriber line
FCC	Federal Communications Commission
FTTH	fiber to the home
HFC	hybrid-fiber coaxial
IEEE	Institute of Electrical and Electronics Engineers
IG	inspector general
ILEC	incumbent local exchange carrier
Kbps	kilobits per second
Mbps	megabits per second
MSA	metropolitan statistical area
NARUC	National Association of Regulatory Utility Commissioners
NATOA	National Association of Telecommunications Officers and Advisors
NCTA	National Cable and Telecommunications Association
NTCA	National Telecommunications Cooperative Association
NTIA	National Telecommunications and Information Administration
RUS	Rural Utilities Service
SIA	Satellite Industry Association
UNE	unbundled network element
USF	Universal Service Fund
USIIA	US Internet Industry Association
USTA	United States Telecom Association
VoIP	voice over Internet protocol
Wi-Fi	wireless fidelity
WiMAX	Worldwide Interoperability for Microwave Access
WISP	wireless Internet service provider
WISPA	Wireless Internet Service Providers Association

May 5, 2006
The Honorable Ted Stevens
Chairman
The Honorable Daniel K. Inouye
Co-Chairman
Committee on Commerce, Science, and Transportation United States Senate

The Honorable Joe L. Barton
Chairman
The Honorable John D. Dingell

Ranking Minority Member
Committee on Energy and Commerce House of Representatives

The universal availability of high speed Internet access over broadband technologies—commonly referred to as broadband Internet access—has become a national goal.[1] The Telecommunications Act of 1996 directed the Federal Communications Commission (FCC) and state commissions to encourage the deployment of advanced telecommunications capability. Similarly, in 2004, the President stated that there should be a national goal for universal, affordable access to broadband technology by 2007. The importance placed on access to broadband correlates to its many benefits for individuals and society. Broadband is seen as a critical economic engine, a vehicle for enhanced learning and medicine, and a central component of 21st century news and entertainment.

As part of our response to a mandate included in the Internet Tax Nondiscrimination Act of 2004, this report examines the factors that affect the deployment—that is, the building of infrastructure over which broadband services can be provided—and the adoption of broadband services. We focus particularly on the deployment and adoption of broadband to households, as opposed to businesses or institutions. In particular, this report provides information on (1) the current status of broadband deployment and adoption; (2) the factors that influence the deployment of broadband networks; (3) the factors that influence the adoption, or purchase, of broadband service by households; and (4) the options that have been suggested to spur greater broadband deployment and adoption. In January 2006, we released a report that examined the impact of the Internet tax moratorium on state and local tax revenues, as also mandated by the law.[2]

To respond to the objectives of this report, we selected eight states and conducted case studies on the status of broadband deployment and adoption. For each of the states—Alaska, California, Kentucky, Massachusetts, North Dakota, Ohio, Texas, and Virginia—we interviewed state and local officials, including local franchising authorities, state public utility regulators, and representatives from governors' offices; state industry and government associations; private cable and telephone providers; wireless Internet service providers; and municipal and cooperative telecommunications providers. We also spoke with a variety of individuals and organizations knowledgeable about broadband services, such as national industry associations and experts. We spoke with representatives from FCC, the National Telecommunications and Information Administration of the Department of Commerce, and the Rural Utilities Service (RUS) of the Department of Agriculture. To assess the status of broadband deployment and to understand the factors affecting the deployment and adoption of broadband, we used survey data from Knowledge Networks/SRI's *The Home Technology Monitor™: Spring 2005 Ownership and Trend Report.* Knowledge Networks/SRI interviewed approximately 1,500 randomly sampled households, asking questions about each household's purchase of Internet services and the availability of cable television service. Using these data, we estimated two econometric models: One model examined the factors affecting broadband deployment and the second examined the factors affecting households' adoption of broadband services. We combined the household survey data with information from FCC's Form 477 filings, which contain information on companies' provision of broadband services by zip codes. This enabled us to develop information about what options for broadband a particular household would have. To assess the impact of Internet taxes on broadband deployment and adoption, we contacted officials in 48 states and the District of

Columbia to determine whether the state, or local governments in the state, imposed taxes on Internet access in 2005; we did not evaluate the level of taxation. We concluded that information from Knowledge Networks/SRI and FCC (with modifications discussed later in this report) was sufficiently reliable for the purpose of this report. All percentage estimates from the Knowledge Networks/SRI survey have margins of error of plus or minus 7 percentage points or less, unless otherwise noted. See appendix I for a more detailed discussion of the overall scope and methodology for this report, including a discussion of how we selected the case-study states; appendix II for an assessment of the data reliability of the Knowledge Networks/SRI survey; and appendix III for a more detailed explanation of, and results from, our deployment and adoption models.

We conducted our work from April 2005 through February 2006 in accordance with generally accepted government auditing standards.

RESULTS IN BRIEF

About 30 million American households purchase, or have adopted, broadband service, but it is difficult to assess the extent of gaps in the availability of broadband in local markets. Using a survey of American households, we found that 28 percent—or about 30 million—subscribed to broadband service in 2005. In addition, 30 percent of surveyed households subscribed to a dial-up Internet service, and 41 percent did not access the Internet from their home. Among households subscribing to broadband service, we found roughly an equal share taking cable modem and digital subscriber line (DSL) service, the two primary broadband services at this time. Households in rural areas were less likely to subscribe to broadband service, compared with households in urban and suburban areas. On a semiannual basis, FCC conducts an extensive data collection effort using its Form 477 to assess the availability of advanced telecommunications service in the United States. As of July 2005, FCC has found that 99 percent of Americans live in the 95 percent of zip codes that have at least one broadband provider reporting to be serving at least one subscriber. These data clearly indicate that deployment of broadband networks has been extensive. However, for its zip-code level data, FCC collects data based on where *subscribers* are served, not where providers have deployed broadband infrastructure. Based on our analysis is appears that these data may not provide a highly accurate depiction of deployment of broadband infrastructures for residential service in some areas.[3]

A variety of market and technical factors, as well as federal and state government efforts and access to resources at the local level have influenced the deployment of broadband infrastructure. Most importantly, companies contemplating the deployment of broadband infrastructure consider both the cost to deploy and operate a broadband network and the expected demand for broadband service. We found it is more costly to serve areas with low population density and rugged terrain with terrestrial facilities than it is to serve areas that are densely populated and have flat terrain. It also may be more costly to serve locations that are a significant distance from a major city. As such, these important factors have caused deployment to be less developed in more rural parts of the country. Firms also consider the extent of existing competition in the broadband market when making deployment decisions: New entrants are more likely to enter markets with no competitors, but at the same time, we found that incumbent cable and telephone companies may respond to entry by new companies by rolling out broadband in markets where they had not yet provided service. Even when cost and demand factors are favorable, technical factors can limit the deployment of broadband service in certain contexts. For example, DSL—the primary broadband service provided by telephone companies—can generally extend only 3 miles[4] from the central office with copper plant, which precludes many households from obtaining DSL service.[5] Finally, we found that a variety of federal and state government efforts as well as access to resources at the local level have influenced the deployment of broadband infrastructure. At the federal level, one of the programs of the Universal Service Fund (USF)—known as the High Cost Fund—has indirectly facilitated broadband service in more rural areas. Similarly, the Department of Agriculture's Rural Utilities Service (RUS) provides grants and loans to promote broadband service in rural areas. At the local level, access to rights-of-way, pole attachments, wireless-tower sites as well as the video franchising process can influence the pace of deployment. We also found that strong leadership within a community can help promote broadband deployment by, for example, enhancing the likely market success of companies' entry into rural markets. Finally, using our econometric model, we found that the imposition of taxes was not a statistically significant factor influencing the deployment of broadband.

A variety of characteristics related to households and services influence whether consumers purchase (or adopt) broadband service. Based on our econometric model, we found that several characteristics of households influence the adoption decision. Our model showed that households with high incomes were 39 percentage points more likely to adopt broadband than lower-income households, and those with a college-educated head of household were 12 percentage points more likely to purchase broadband than households headed by someone who did not graduate from college. While rural households are less likely to adopt broadband, our findings indicate that this difference may be related in part to the lower availability of broadband in rural areas. In addition, based on discussions with stakeholders, we identified several characteristics of broadband service that influence whether a consumer purchases the service. The price of broadband service remains a barrier to adoption of broadband service for some consumers, although prices have been declining recently. The availability of applications and services that either require or function much more effectively with broadband—such as computer gaming and file sharing—also influences whether a particular consumer purchases broadband service. Using our model, we found that the imposition of the tax was not a statistically significant factor influencing the adoption of broadband service at the 5 percent level. It was statistically significant at the 10 percent level, perhaps suggesting that it is a weakly significant factor. However, giving the nature of our model, it is unclear whether this finding is related to the tax or other characteristics of the states in which the households resided.

Targeted government assistance might help facilitate the deployment of broadband service, and stakeholders we spoke with identified several options to spur greater deployment of broadband service in rural America. However, each of the policy options that stakeholders discussed with us had challenges to their implementation. For example, a few of the stakeholders we spoke with expressed concerns about the structure of the Rural Utilities Service's broadband loan program. Also, several of the stakeholders suggested that modifications to spectrum management might address the lack of broadband infrastructure in rural areas. Finally, because the cost of building land-based infrastructure is so high in some rural areas, satellite industry stakeholders noted that satellite broadband technology may be the best option for addressing a lack of broadband in those regions. Ultimately, we found that a key difficulty for analyzing and targeting any federal aid for broadband is a lack of reliable data on the deployment of networks.

We provided a draft of this report to the Department of Agriculture, the Department of Commerce, and FCC for their review and comment. The Department of Agriculture had no comments on the draft. The Department of Commerce and FCC provided technical comments that we incorporated, as appropriate.

In the draft, GAO recommended that FCC identify and evaluate strategies for improving the 477 data such that the data provide a more accurate depiction of residential broadband deployment throughout the country. In oral comments regarding this recommendation, FCC staff noted that the commission had recently determined that it would be costly and could impose large burdens on filers—particularly small entities—to require any more detailed filings on broadband deployment. As such, we recommend that FCC develop information regarding the degree of cost and burden that would be associated with various options for improving the information available on broadband deployment and should provide that information to the Senate Committee on Commerce, Science, and Transportation and the House Energy and Commerce Committee in order to help them determine what actions, if any, are necessary going forward. FCC did not comment on our final recommendation.

We also provided a draft of this report to several associations representing industry trade groups and state and local government entities for their review and comment. Specifically, the following associations came to GAO headquarters to review the draft: Cellular Telecommunications and Internet Association (CTIA), National Association of Regulatory Utility Commissioners (NARUC), National Association of Telecommunications Officers and Advisors (NATOA), National Cable and Telecommunications Association (NCTA), National Telecommunications Cooperative Association (NTCA), Satellite Industry Association (SIA), US Internet Industry Association (USIIA), United States Telecom Association (USTA), and Wireless Internet Service Providers Association (WISPA). Officials from CTIA, NARUC, and NTCA did not provide comments. Officials from NATOA, NCTA, SIA, and USIIA provided technical comments that were incorporated, as appropriate. USTA and WISPA provided comments that are discussed in appendix V.

BACKGROUND

Internet access became widely available to residential users by the mid 1990s. For a few years, the primary mechanism to access the Internet was a dial-up connection, in which a standard telephone line is used to make an Internet connection. A dial-up connection offers data transmission speeds up to 56 kilobits per second (Kbps). Broadband, or high-speed, Internet access became available by the late 1990s. Broadband differs from a dial-up connection in certain important ways. First, broadband connections offer a higher-speed Internet connection than dial-up—for example, some broadband connections offer speeds exceeding 1 million bits per second (Mbps) both upstream (data transferred from the consumer to the Internet service provider) and downstream (data transferred from the Internet service provider to the consumer).[6] These higher speeds enable consumers to receive information much faster and thus enable certain applications to be used and content to be accessed that might not be possible with a dial- up connection. Second, broadband provides an "always on" connection to the Internet, so users do not need to establish a connection to the Internet service provider each time they want to go online.

Consumers can receive a broadband connection to the Internet through a variety of technologies. These technologies include, but are not limited to, the following:

Cable modem. Cable television companies first began providing broadband service in the late 1990s over their hybrid-fiber coaxial networks. When provided by a cable company, broadband service is referred to as cable modem service. Cable providers were upgrading their infrastructure at that time to increase their capacity to provide video channels in response to competition from direct broadcast satellite (DBS) providers such as DirecTV® and Dish Network. By also redesigning their networks to provide for two-way data transmission, cable providers were able to use their systems to provide cable modem service. Cable modem service is primarily available in residential areas, and although the speed of service varies with many factors, download speeds of up to 6 Mbps are typical. Cable providers are developing even higher speed services.

DSL. Local telephone companies provide digital subscriber line (DSL) service, another form of broadband service, over their telephone networks on capacity unused by traditional voice service. Local telephone companies began to deploy DSL service in the late 1990s—some believe, in part, as a response to the rollout of cable modem service. To provide DSL service, telephone companies must install equipment in their facilities and remove devices on phone lines that may cause interference. While most residential customers receive asymmetric DSL (ADSL) service with download speeds of 1.5 to 3 Mbps, ADSL technology can achieve speeds of up to 8 Mbps over short distances. Newer DSL technologies can support services with much higher download speeds.

Satellite. Currently, three providers of satellite service can offer nearly ubiquitous broadband service in the United States. These providers use geosynchronous satellites that orbit in a fixed position above the equator and transmit and receive data directly to and from subscribers. Signals from satellites providing broadband service can be accessed as long as the user's reception dish has a clear view of the southern sky. Therefore, while the footprint of the providers' transmission covers most of the country, a person living in an apartment with windows only facing north, or a person living in house in a heavily wooded area might not be able to receive Internet access via satellite. Earlier Internet services via satellite could only receive Internet traffic downstream— that is, from the satellite to the subscriber—and upstream Internet traffic was transmitted through a standard telephone line connection. Currently, however, satellite companies provide both upstream and downstream connections via satellite, eliminating the need for a telephone line connection and speeding the overall rate of service. Transmission of data via satellite typically adds one-half to three-fourths of a second, causing a slight lag in transmission and rendering this service less well-suited for certain applications over the Internet. While satellite broadband service may be available throughout the country, the price for this service is generally higher than most other broadband modes; both the equipment necessary for service and recurring monthly fees are generally higher for satellite broadband service, compared with most other broadband transmission modes.

Wireless. Land-based, or terrestrial, wireless networks can offer a broadband connection to the Internet from a wide variety of locations and in a variety of ways. Some services are provided over unlicensed spectrum and others over spectrum that has been licensed to particular companies.[7] In licensed bands, some companies are offering fixed wireless broadband throughout cities. Also, mobile telephone carriers—such as the large companies that provide traditional cell phone service—have begun offering broadband mobile wireless Internet service over licensed spectrum—a service that allows subscribers to access the Internet with their mobile phones or laptops as they travel across cities where their provider supports the service. Such services are becoming widely deployed and are increasingly able to offer high-speed services. A variety of broadband access technologies and services are also provided on unlicensed spectrum— that is, spectrum that is not specifically under license for a particular provider's network. For example, wireless Internet service providers generally offer broadband access in particular areas by placing a network of antennae that relay signals throughout the network. Subscribers place necessary reception equipment outside their homes that will transmit and receive signals from the nearest antenna. Also, wireless fidelity (Wi-Fi) networks—which provide broadband service in so-called "hot spots," or areas up to 300 feet—can be found in cafes, hotels, airports, and offices. Some technologies, such as Worldwide Interoperability for Microwave Access (WiMAX), can operate on either licensed or unlicensed bands, and can provide broadband service up to approximately 30 miles in a line-of-sight environment.

Under section 706 of the Telecommunications Act of 1996, Congress directs FCC to encourage deployment of advanced telecommunications capability, which includes broadband, to all Americans. In implementing the act, FCC has treated the two most widely available broadband services—cable modem and DSL service—as information services having a telecommunications component. FCC's approach of not treating such services as telecommunications services has important legal implications because a service defined as a telecommunications service could be subject to regulation under Title II of the Communications Act, which imposes substantial common carrier regulations unless the commission choose to forebear from their enforcement. As part of its responsibilities, FCC periodically issues a report to Congress on the status of advanced telecommunication capability in the United States. To prepare this report,

FCC developed a periodic reporting requirement using Form 477. In November 2004, FCC modified its rules regarding the filing of the 477 form, which went into effect for the companies' second filing in 2005. Specifically, FCC removed existing reporting thresholds,[8] and companies are now required to report their total state subscribership by technology.[9]

About 30 Million American Households Purchase Broadband Service; Despite Evidence of Substantial Broadband Deployment throughout the United States, It Is Difficult to Assess Deployment Gaps in Some Areas

We found that in 2005, about 30 million American households—or 28 percent—subscribed to broadband, although households in rural areas were less likely to subscribe to broadband service than were households in urban and suburban areas. Households appear to subscribe to cable modem and DSL services—the two primary broadband services—in approximately equal numbers. FCC requires providers of broadband service to report on the geographic areas in which they serve *subscribers*, but these data are sometimes used to infer the status of *deployment* of companies' Internet infrastructure. Some stakeholders find FCC data collection efforts useful for comparison of adoption of broadband across states, but we found that the data may not be as useful for understanding the status of broadband deployment across the country.

About 30 Million American Households Purchase Broadband Service

Based on survey data from 2005,[10] we found that 28 percent of American households subscribe to broadband service. Figure 1 illustrates how American households access the Internet, by various technologies, and also shows the percentage of households that do not own a computer. As shown, 30 percent of American households subscribe to dial-up access, and about 41 percent of American households do not have an Internet connection from home. Of those households that do not access the Internet, more than 75 percent do not have a computer in the home, while the remaining households own a computer but do not have online access.

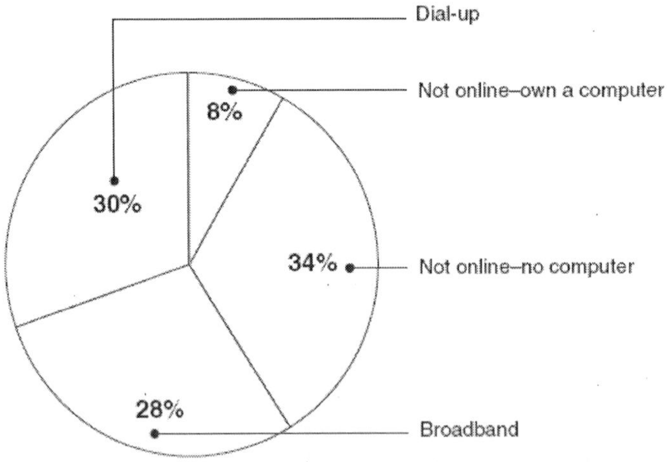

Source: GAO analysis of Knowledge Networks/SRI's The Home Technology MonitorTM: Spring 2005 Ownership and Trend Report.

Figure 1. Status of Household Computer Ownership and Internet Connection.

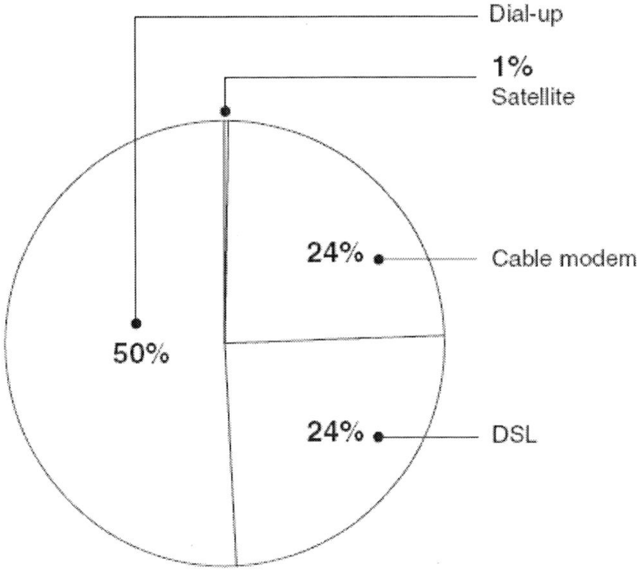

Source: GAO analysis of Knowledge Networks/SRI's The Home Technology MonitorTM: Spring 2005 Ownership and Trend Report.

Figure 2. Household Online Connection.

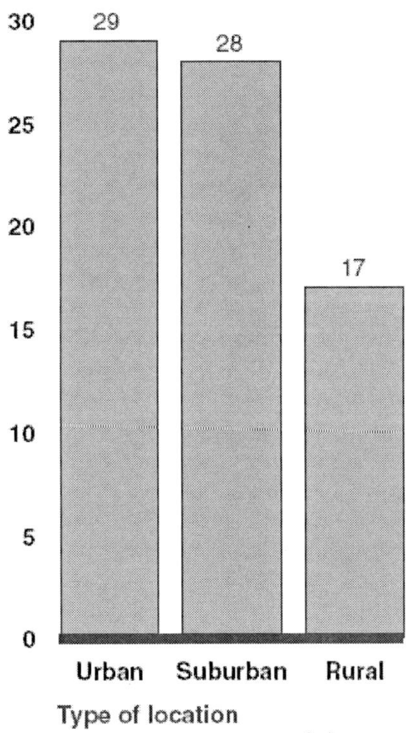

Source: GAO analysis of Knowledge Networks/SRI's The Home Technology MonitorTM: Spring 2005 Ownership and Trend Report.

Figure 3. Percentage of Households Subscribing to Broadband, by Type of Location.

Among online households, we found 50 percent subscribe to dial-up service, and 48 percent subscribe to a broadband service.11 Additionally, we found that of those households subscribing to a broadband service, roughly half purchase DSL service and half purchase cable modem service. (See fig. 2 for the types of connections purchased by online households.)

Finally, we found that households residing in rural areas were less likely to subscribe to broadband service than were households residing in suburban and urban areas.12 Seventeen percent of rural households subscribe to broadband service, while 28 percent of suburban and 29 percent of urban households subscribe to broadband service. (See fig. 3 for the percentage of urban, suburban, and rural households purchasing broadband service.)

We also found that rural households were slightly less likely to connect to the Internet, compared with their counterparts in suburban areas.[13]

Deployment of Broadband Appears to Be Extensive, but FCC's Form 477 Data May not Provide an Accurate Depiction of Gaps in Broadband Deployment

In order to fulfill its responsibility under section 706 of the Telecommunications Act, FCC collects data on companies' broadband operations. In early 2004, FCC initiated a proceeding to examine whether it should collect more detailed information for its broadband data gathering program than had previously been collected. Specifically, FCC asked whether it should do several things to enhance the broadband data including (1) requiring providers to report the speeds of their broadband services, (2) eliminating the reporting threshold such that all providers of broadband—no matter how small—must report to FCC on its services, and (3) requiring that providers report the number of connections by zip code. In late 2004, FCC released an order in which it decided to require all providers—no matter how small—of broadband to report in the 477 data collection effort on broadband and also required providers to report information about their services within speed tier categories. The commission decided not to require providers to report the number of connections (or subscribers) that they serve within each zip code or the number of connections in speed tiers or by technology within each zip code, finding that finding that such a requirement would impose a large burden on filers (particularly smaller entities), and would require significant time to implement. In particular, several providers commented in the 2004 proceeding that it would be costly and burdensome to develop the software and systems to generate the detailed zip code-level data and that the cost and burden of further reporting requirements would likely outweigh the benefits of more substantial information on broadband deployment in the United States. On the other hand, 3 state utility commissions asked FCC to gather more information within zip codes or by some other Census boundary because such information is, in their view, important for tracking broadband availability.

Based on the modifications to the filing requirements FCC implemented, FCC collects, through its Form 477 filings, information on several aspects of each company's provision of broadband at the state level, such as the total number of subscribers served, the breakdown of how those subscribers are served by technology, and estimates within each technology of the percentage of subscribers that are residential. For each technology identified in the state reporting, providers also submit a list of the zip codes in which they serve at least one customer. As discussed above, companies do not report the number of subscribers served or whether subscribers are business or residential within each zip code; they also do not report information on the locations within the zip code that they *can* serve.

In July 2005, FCC found that 99 percent of the country's population lives in the 95 percent of zip codes where at least one provider reported to FCC that it serves at least one high-speed subscriber as of December 31, 2004. In 83 percent of the nation's zip codes, FCC noted that subscribers are served by more than one provider, and the commission noted that for roughly 40 percent of zip codes in the United States, there are five or more providers reporting high-speed lines in service. Although these data indicate that broadband availability is extensive, we found that FCC's 477 data may not be useful for assessing broadband deployment at the local level.[14] While FCC, in general, notes that the 477 zip-code data are not meant to measure deployment of broadband, in its July 2005 report,[15] the commission states that in order to be able to evaluate *deployment*, the commission "instituted a formal data collection program to gather standardized information about *subscribership to high speed services. . . .*" (Emphasis added.) Based on our analysis, we found that collecting data about where companies have *subscribers* may not provide a clear depiction of their *deployment*, particularly in the context of understanding the availability of broadband for *residential* users.[16]

One quandary in analyzing broadband deployment is how to consider the availability of satellite broadband services. Even though broadband over satellite may not be seen by some as highly substitutable for other broadband technologies because of certain technical characteristics or because of its higher cost, satellite broadband service is *deployed*: Three companies have infrastructure in place to provide service to most of the country.[17] The actual *purchase* of satellite broadband is scattered throughout the country and shows up in FCC's 477 zip-code data only where someone actually purchases the service. It is not clear how satellite service should be judged in terms of deployment. Since it is available throughout the entire country, one view could be that broadband is near fully deployed. Alternatively, it could be viewed that satellite broadband— while available in most areas—does not reflect *localized* deployment of broadband infrastructure and should therefore not be counted as a deployed broadband option at all. In either case, FCC's zip-code data on satellite broadband—which is based on the pattern of scattered subscribership to this service—does not seem to be an appropriate indicator of its availability.

Aside from the question of how to view satellite deployment, other issues arise in using subscribership indicators for wire or wireless land-based providers at the zip-code level as an indicator of deployment. These issues include the following:

- Because a company will report service in a zip code if it serves just one person or one institution in that zip code, stakeholders told us that this method may overstate deployment in the sense that it can be taken to imply that there is deployment throughout the zip code even if deployment is very localized. We were told this issue might particularly occur in rural areas where zip codes generally cover a large geographic area. Based on our own analysis, we found, for example, that in some zip codes more than one of the large established cable companies reported service. Because such providers rarely have overlapping service territories, this likely indicates that their deployment was not zip-code-wide and that the number of providers reported in the zip code overstates the level of competition to individual households. We also found that a nontrivial percentage of households lie beyond the 3-mile radius of their telephone central office, indicating that DSL service was unlikely to be available to these homes.

- Companies report service in a zip code even if they only serve businesses. One academic expert we interviewed expressed a concern about this issue. Based on our own analysis, we found that many of the companies filing 477 data indicating service in particular zip codes only served business customers. As such, the number of providers reported as serving many zip codes is likely overstated in terms of the availability of broadband to residences.
- FCC requires that companies providing broadband using unbundled network elements (UNE)[18] report their broadband service in the zip code data. When a provider serves customers using UNEs, they purchase or lease underlying telecommunications facilities from other providers—usually incumbent telephone companies—to serve their customers. Having these providers report their subscribers at the state level is important to ensure that correct numbers on the total subscribers of broadband service is obtained. However, while UNE providers may make investments in infrastructure, such as in collocation equipment, they do not generally own or provide last-mile connectivity for Internet access. Thus, counting these providers in the zip-code-level data may overstate the extent of local infrastructure deployment in the sense that several reporting providers could be relying on the same infrastructure, owned by the incumbent telephone company, to provide broadband access.

ConnectKentucky

Source: ConnectKentucky.

The purpose of ConnectKentucky's Geographic Information Systems (GIS) mapping project is to produce an inventory of existing broadband infrastructure and service availability. The tool can produce maps at the state and census block level. Some of the items mapped include water towers, wireless towers, proposed sewer lines, roads, and population density. The maps also plot which areas are served by municipal, local exchange carrier, cable, and wireless broadband providers.

Based on our analysis, we believe that the use of subscriber indicators at the zip-code level to imply availability, or deployment, may overstate terrestrially based deployment. We were able to check these findings for one state—Kentucky—where ConnectKentucky, a state alliance on broadband, had done an extensive analysis of its broadband deployment. ConnectKentucky officials shared data with us indicating that approximately 77 percent of households in the state had broadband access available as of mid-2005. In contrast, we used population data within all zip codes in Kentucky, along with FCC's 477 zip-code data for that state, and determined that, according to FCC's data, 96 percent of households in Kentucky live in zip codes with broadband service at the end of 2004. Thus, based on the experience in Kentucky, it appears that FCC's data may overstate the availability and competitive deployment of nonsatellite broadband.

Additionally, to prepare our econometric models, we relied on FCC's 477 data to assess the number of providers serving the households responding to Knowledge Networks/SRI's survey. Based on FCC's data, we found that the median number of providers reporting that they serve zip codes where the households were located was 8; in 30 percent of these zip codes, 10 or more providers report that they provide service. Only 1 percent of respondents lived in zip codes for which no broadband providers reported serving at least one subscriber, according to FCC's data. To better reflect the actual number of providers that each of the survey respondents had available at their residence, we made a number of adjustments to FCC's provider count based on our analysis of the providers, certain geographic considerations, and information provided by the survey respondents.[19] After making these adjustments, the median number of providers for the respondents fell to just 2, and we found that 9 percent of respondents likely had no providers of broadband at all.

Despite these concerns about FCC's 477 data, several stakeholders, including a state regulatory office and a state industry association, said they found FCC's data useful. An official at a state governor's office also noted that analysis of FCC data allowed them to make conclusions about the extent of deployment in their state. Similarly, an official in another governor's office said that they use FCC's data to benchmark the accessibility of broadband in their state because it is the only data available. A few academic experts also told us that they use FCC's data.

A Variety of Market and Technical Factors, in Addition to Government Involvement and Access to Resources at the Local Level, have Influenced the Deployment of Broadband

Several market characteristics appear to influence providers' broadband deployment decisions. In particular, factors related to the cost of deploying and providing broadband services, as well as factors related to consumer demand, were critical to companies' decisions about whether to deploy broadband infrastructure. At the same time, certain technical factors related to specific modes of providing broadband service influence how and where this service can be provided. Finally, a variety of federal and state government activities, as well as access to resources at the local level, have influenced the deployment of broadband infrastructure.

Several Key Market Factors Related to the Cost of Service and Demand Influence Deployment Decisions

As companies weigh investment decisions, they consider the likely profitability of their investments. In particular, because broadband deployment requires substantial investment, potential providers evaluate the cost to build and operate the infrastructure, as well as the likely demand—that is, the expected number of customers who will purchase broadband service at a given price—for their service. Based on our interviews, we found that several cost and demand factors influence providers' deployment decisions.

Cost Factors

The most frequently cited cost factor affecting broadband deployment was the population density of a market. Many stakeholders, including broadband providers, state regulators, and state legislators, said population density—which is the population per square mile—was a critical determinant of companies' deployment decisions. In particular, we were told that the cost of building a broadband infrastructure in areas where people live farther apart is much higher than building infrastructure to serve the same number of people in a more urban setting. As such, some stakeholders noted that highly rural areas—which generally have low population density—can be costly to serve. Results from our econometric model confirm the views of these stakeholders. We found that densely populated and more urbanized locations were more likely to receive broadband service than were less densely populated and rural locations. For example, we found that urban areas were 9 percentage points more likely to have broadband service available than were rural areas.

Source: GAO.

In Alaska, backhaul from rural villages requires the use of satellites. This type of backhaul is costly because of the need for terrestrial infrastructure to send and receive signals from satellites as well as the need to either own or lease satellite transmitters. The high cost can affect whether providers deploy broadband service in a village. To help defray this cost, providers often look to serve an "anchor tenant" in a village, such as schools or health clinics that receive federal funding.

Terrain was also frequently cited as a factor affecting broadband deployment decisions. In particular, we were told that infrastructure build- out can be difficult in mountainous and forested areas because these areas may be difficult to reach or difficult on which to deploy the required equipment. Conversely, we were told that flat terrain constitutes good geography for telecommunications deployment. For wireless providers, we were told that terrain concerns can present particular challenges because some wireless technologies require "line-of-sight," meaning that radio signals transmitted from towers or antennas need an unobstructed pathway—with no mountains, trees, or buildings—from the transmission site to the reception devices at users' premises. Terrain can also affect satellite broadband availability in rural areas that have rolling hills or many trees that can obstruct a satellite's signal.

Some stakeholders also said costs for what is known as "backhaul" are higher for rural areas and can affect the deployment of broadband networks in these areas. Backhaul refers to the transmission of information—or data—from any of a company's aggregation points to an Internet backbone provider that will then transmit that data to any point on the Internet. This is also sometimes referred to as the "middle mile." Internet traffic originating from rural areas may need to travel a long distance to a larger city to connect to a major Internet backbone provider. Because the cost of transmitting over this distance—that is, the backhaul— can be high, one stakeholder noted that backhaul costs are another barrier to deployment in rural areas. However, using our econometric model, we did not find that the distance to a metropolitan area with a population of 250,000 or more—our proxy for backhaul—was associated with a lower likelihood of broadband deployment.

Demand Factors

Based on our interviews with stakeholders, we found that certain demand factors influence providers' deployment decisions. In particular, because stakeholders noted that potential providers seek to deploy in markets where demand for their service will be sufficient to yield substantial revenues, certain elements of markets were identified as affecting the demand for broadband:

- *Ability to aggregate demand.* For rural locations, officials we spoke with stressed the importance of aggregating sufficient demand. For example, officials in one state told us that to justify the cost of deployment in rural areas where population density is low, telecommunications providers need to be able to aggregate all of the possible demand to make a business case. We were also told that aggregation is furthered by ensuring that a large "anchor tenant" will subscribe to the service. Possible anchor-tenant customers described to us included large businesses, government agencies, health-care facilities, and schools. Because the revenues from such customers will be significant, two stakeholders noted that the anchor tenant alone will help to cover a significant portion of the providers' expenses.
- *Degree of competition.* We found that the degree of existing broadband competition in a local market can inhibit or encourage deployment, depending on the circumstances. Some new entrants—companies not already providing a telecommunications service in an area—reported that they avoid entering markets with several existing providers and seek out markets where incumbent telephone and cable companies do not provide broadband service. The lack of existing service

enables the entrant company to have the potential to capture many customers. At the same time, stakeholders told us that deployment by a new entrant often spurred incumbent telephone or cable providers to upgrade their infrastructures so as to compete with the entrant in the broadband market.

- *Technological expertise.* A few stakeholders noted that demand will be greater in areas where potential customers are familiar with computers and broadband, as these individuals are more likely to purchase broadband service.

Stakeholders we spoke with rarely mentioned the per-capita income of a service area as a factor determining deployment. In fact, a few stakeholders credited cable franchising requirements with ensuring deployment to low- income areas; in some cases, cable franchise agreements require cable providers to build out to all parts of the service territory. However, a 2004 study did find that areas with higher median incomes were more likely to have competitive broadband systems.[20] Similarly, results from our econometric analysis indicate that areas with higher per-capita income are more likely to receive broadband service than are areas with lower per- capita income.

Taxation of Internet Access

Using our econometric model, we did not find that taxation of Internet access by state governments influenced the deployment of broadband service. Taxes can raise consumer prices and reduce revenues and impose costs on providers, and thereby possibly reduce the incentive for companies to deliver a product or service. To assess the impact of Internet taxes on broadband deployment, we contacted officials in 48 states and the District of Columbia[21] to determine whether the state, or local governments in the state, imposed taxes on Internet access. To incorporate this analysis into our model, we used a binary variable to indicate the presence of the tax; that is, each state was placed into one of two groups, states with a tax and states without a tax. As such, this binary variable could potentially capture the influence of other characteristics of the states, in addition to the influence of the tax. While the parameter estimate in our model had the expected sign—indicating that the imposition of taxes may reduce the likelihood of broadband deployment—it was not statistically significant.

Certain Critical Technical Factors Affect Broadband Deployment

Many stakeholders we spoke with commented on issues related to technical characteristics of networks that provide broadband. In particular, many noted that certain technical characteristics of methods used to deliver broadband influence the extent of its availability. In terms of issues discussed for established modes of broadband delivery, we were told the following:

- DSL service can generally be provided over telephone companies' copper plant to residences and businesses that are within approximately 3 miles from the telephone company's facility, known as a central office. However, if the quality of the telephone line is not good, the distance limit can be reduced—that is, it may only be possible to provide DSL for locations within some lesser distance—perhaps 2 miles—from a central office. We were told, for example, that in some rural areas, DSL is more limited by distance because the telephone lines may be older. While the distance limit of DSL can be addressed by deploying certain additional equipment that extends fiber further into the network, this process can be expensive and time consuming.

- Across spectrum bands used to provide terrestrial wireless broadband service, technical characteristics vary: In some cases, signals may travel only a short distance, and in other cases, they may travel across an entire city; in some cases there may be a need for line-of-sight from the transmission tower to the user, but in other cases, the signals may be able to travel through walls and trees. Some stakeholders mentioned that wireless methods hold great promise for supporting broadband service.
- Satellite technology can provide a high-speed Internet service throughout most of the United States. However, the most economical package of satellite broadband service generally offers, at this time, upstream speeds of less than 200 kilobits per second, and therefore this service does not necessarily meet FCC's definition of *advanced telecommunications services*, while it does meet FCC's definition of high-speed service. Despite the near universal coverage of satellite service, consumers need a clear view of the southern sky to be able to receive transmissions from the satellites. Additionally, transmission via satellite introduces a slight delay, which causes certain applications, such as VoIP (i.e., telephone service over the Internet), and certain computer gaming to be ill-suited for use over satellite broadband.

Some emerging or expanding broadband technologies do not currently have significant subscribership, but have the potential to be important means of broadband service in the coming years. These technologies include deep fiber deployment (e.g., fiber to the home), WiMAX, broadband over power lines (BPL), and third-generation (3G) cellular. Each of these technologies has technical considerations that are influencing the nature of deployment. See appendix IV for a discussion of these technologies.

Federal and State Government Efforts, and Access to Resources at the Local Level, have Impacted the Deployment of Broadband

We found that government involvement in several venues, and access to resources at the local level, have affected the deployment of broadband networks throughout the nation. In particular, we found that (1) certain federal programs have provided funding for broadband networks; (2) some state programs have assisted deployment; (3) state and local government policies, as well as access to resources at the local level, can influence broadband deployment; and (4) broadband deployment—particularly in more rural settings—is often influenced by the extent of involvement and leadership exercised by local government and community officials.

Federal Programs have Funded Broadband Infrastructure

We found that several federal programs have provided significant financial assistance for broadband infrastructure.

- The Universal Service Fund (USF) has programs to support improved telecommunications services. The high-cost program of the USF provides eligible local telephone companies with funds to serve customers in remote or rural areas where the cost of providing service is higher than the cost of service in more urbanized areas. The high-cost funds are distributed to providers according to formulas based on several factors, such as the cost of providing service, with funds distributed to small rural incumbent local exchange carriers (ILEC) and larger ILECs serving rural areas based on different formulas. Competitive local exchange carriers can also qualify to receive high-cost funds. While high-cost funds are not specifically targeted to support the deployment of broadband infrastructure, these funds do support *telecommunications* infrastructure that is also used to provide broadband services. We were told by some stakeholders in certain states that high-cost support has been very important for the upgrade of telecommunications networks and the provision of broadband services. In particular, some stakeholders in Alaska, Ohio, and North Dakota told us that high-cost support has been critical to small telephone companies' ability to upgrade networks and provide broadband services. Additionally, the e-rate program of the USF has provided billions of dollars in support of Internet connectivity for schools and libraries. Another USF program, the Rural Health Care Program, provides assistance for rural health facilities' telecommunications services.
- Some programs of the U.S. Department of Agriculture's Rural Utilities Service (RUS) provide grants to improve rural infrastructures providing broadband service. The Community Connect Program provides grants to deploy transmission infrastructures to provide broadband service in communities where no broadband services exist, and requires grantees to wire specific community facilities and provide free access to broadband services in those facilities for at least 2 years. Grants can be awarded to entities that want to serve a rural area of fewer than 20,000 residents. Approximately $9 million was appropriated in 2004 as well as in 2005 for this purpose.
- RUS's Rural Broadband Access Loan and Loan Guarantee program provides loans[22] to eligible applicants to deploy infrastructures that provide broadband service in rural communities that meet the program's eligibility requirements. A wide variety of entities are eligible to obtain loans to serve small rural communities. To obtain a 4 percent loan, the applicant must plan on serving a community with no previously available broadband service, but loans at the Treasury interest rate do not have such a requirement.
- The Appalachian Regional Commission's Information Age Appalachia program focuses on assisting in the development and use of telecommunications infrastructure. The program also provides funding to assist in education and training, e-commerce readiness, and technology-sector job creation. We were told that in Kentucky, funding from the commission assisted the development and operations of

ConnectKentucky, a state alliance that focuses on broadband deployment and adoption. The Appalachian Regional Commission also provided some funding for projects in Ohio and Virginia.

Universal Service Fund

Source: GAO.

The Universal Service Fund facilitates education and health care in rural Alaska by providing broadband Internet connections. Access to broadband connections provides learning opportunities for children, such as higher-level curriculum, and teacher certification programs. Additionally, it facilitates health care by allowing more immediate care in local villages rather than requiring long trips to a larger clinic or hospital.

Various State Programs Assist the Deployment of Broadband Services

A number of states we visited have had programs to assist the deployment of broadband services, including the following:

- The Texas Telecommunications Infrastructure Fund began in 1996 and according to an official of the Texas Public Utility Commission committed to spend $1 billion on telecommunications infrastructure in Texas. Public libraries, schools, nonprofit medical facilities, and higher education institutions received grants for infrastructure and connectivity to advanced communications technology. The program is no longer operational.
- ConnectKentucky is an alliance of technology-focused businesses, government entities, and universities that work together to accelerate broadband deployment in the state. ConnectKentucky focuses on three goals: (1) raising public awareness of broadband services, (2) creating market-driven strategies to increase demand—particularly in rural areas, and (3) initiating policy to reduce regulatory barriers to broadband deployment.
- The Virginia Tobacco Indemnification and Community Revitalization Commission partially funded Virginia's Regional Backbone Initiative. The backbone initiative is designed to stimulate economic development opportunities by encouraging the creation of new technology-based business and industry in southern Virginia, which has historically relied heavily on tobacco production.

Local Issues and Access to Resources Impact the Deployment of Broadband Services

The ability of a company to access local rights-of-way, telephone and electric poles, and wireless-tower sites can influence the deployment of broadband service. In particular, a few stakeholders we spoke with said difficulty in gaining access to these resources can serve as a barrier to the rapid deployment of broadband service because accessing these resources was a time-consuming and expensive process. Companies often require access to rights-of-way—such as areas along public roads—in order to install infrastructure for broadband service. In some instances, companies can face challenges gaining access to rights-of-way, which can hinder broadband deployment. For example, we were told that in one California community, providers had difficulty bringing wires across a highway, which limited their ability to provide service in all areas of the community. Some companies also require access to telephone and electric poles to install their broadband infrastructure. Depending on the entity owning the pole, we were told that acquiring access to poles could be costly and time consuming. For example, one BPL provider told us that it encountered difficulty accessing poles owned by the telephone company. Finally, wireless companies need access to towers or sites on which they can install facilities for their broadband infrastructure. A few stakeholders we spoke with told us that gaining this access can be a difficult process. For example, one company said providers are often challenged by the need to learn each town's tower-siting rules. While some stakeholders identified problems gaining access to these resources, other stakeholders did not identify access to rights-of-way, poles, and other resources as issues in deploying broadband services.

We found that the video-franchising process can also influence the deployment of broadband service because companies may be building infrastructure to simultaneously provide both video and broadband services. To provide video service, such as cable television, companies usually must obtain a franchise agreement from a local government. Some stakeholders assert that the video-franchising process can delay the deployment of broadband service because providers must negotiate with a large number of local jurisdictions. Further, these negotiations can be time consuming and costly. As a result, these stakeholders believe that local franchising can hinder their ability to deploy broadband infrastructures. Alternatively, some stakeholders believe that the video-franchising process is important because it helps promote deployment of broadband service to all areas of a community. For example, some jurisdictions have ubiquity requirements mandating deployment to all areas of a community, including those that are less affluent. These stakeholders argue that without the local ubiquity requirement, service providers could "cherry pick" and exclusively provide broadband services to more economically desirable areas.

In some instances, municipal governments provide broadband infrastructure and service. For example, we spoke with officials in five municipal governments that provide wire-based broadband service, often in conjunction with the government's electric utility. We also spoke with one municipal government that provided wireless broadband service. A few of these municipal government officials told us that their municipality had undertaken this deployment because they believe that their communities either do not have, or do not have adequate, private broadband service. A significant number of stakeholders we interviewed support a municipality's right to provide broadband services and believe that broadband service is a public utility, such as water and sewer. Some support municipal deployment of broadband, regardless of whether other providers are available in that area, while other stakeholders support a municipality's right to deploy broadband service only if there are no other broadband providers serving the area. However, other stakeholders we spoke with oppose municipal government deployment of broadband service. These stakeholders believe that municipal governments are not prepared to be in the business of providing broadband and that municipal deployment can hinder private-sector deployment.

Community Leadership Encourages the Deployment of Broadband Services

We found that the cost of serving rural areas presents a challenge to the nationwide goal of universal access to broadband. One of the ways that some communities have addressed the lack of market entry into rural areas has been through initiatives wherein community leaders have worked to enhance the likely market success of private providers' entry into rural broadband markets. For example, some community leaders have worked to aggregate demand—that is, to coordinate the Internet needs of various users so that a potential entrant would be able to support a business plan. We were told that this leadership—sometimes by key government officials, sometimes through partnerships—was seen as critical in helping to spur the market in some unserved areas.[23] The following examples illustrate this point:

In Massachusetts, several regional coalitions that have been called "connect" projects focus on demand aggregation as a tool to encourage further deployment of telecommunications backbone and broadband networks in more rural parts of the state that were not well served by broadband providers. In particular, three such regional groups said their demand aggregation model is designed to maximize the purchase of broadband services in their region by working with local hospitals, schools, home businesses, small business, and residents to demonstrate the full extent of the demand for broadband and thus encourage private investment in infrastructure. For the one project that was the most developed, a few stakeholders told us that the group had been critical in helping to spur infrastructure development in the area, and that leadership by State government was important to the development of the initiative.

Berkshire Connect Attracts Deployment to Western Massachusetts

BERKSHIRE CONNECT, INC.

Source: Berkshire Connect, Inc.

Berkshire Connect, Inc. was founded to respond to the lack of reliable and affordable T-1 service for businesses in the Berkshire region of Massachusetts. The organization aggregates demand for broadband service among users in the region, such as hospitals, schools, large and small businesses, and nonprofit organizations. Through a request for proposal process, the organization selected "preferred providers" that offer members of Berkshire Connect an array of advanced telecommunications services at competitive prices. Berkshire Connect's primary goals are to provide and ensure competitive pricing, equal pricing throughout Berkshire County, sustained competition, and a community focus regarding the provision of telecommunications services.

ConnectKentucky, as discussed earlier, is an example of a state coalition taking a leadership role to develop information on state deployment levels, educate citizens about the benefits of broadband service, and advocate broadband-friendly policies with the state legislature. Throughout our meetings in Kentucky, the work of ConnectKentucky was stated to have been instrumental in the development of a common understanding of the state of broadband deployment and adoption as well as in instigating new initiatives to advance the market. The key element of ConnectKentucky that was cited as crucial to its success was leadership from state government, in particular from the governor's office.

In Alaska, we found that in one remote area—Kotzebue, a community 26 miles above the Arctic Circle—strong local leadership was important to the development of a public-private partnership that provides improved medical care to the region. The local leadership from the health cooperative brought together parties in the community and worked with them to develop a plan to provide enhanced health service throughout the community's villages. The Maniilaq Health Center uses a wireless "telecart" with a video camera that can send high-quality, real-time sound and video between the center and Anchorage. The center's physicians are able to perform procedures under the guidance of experts in Anchorage who can "remotely" look over the physicians' shoulders. In addition, there are village clinics staffed by trained village health aides. These village clinics are connected to the main health center via a broadband link that allows them to share records and diagnoses via the telecart.

A Variety of Household and Service Characteristics Influence the Adoption of Broadband

We developed an econometric model to assess the many factors that might influence whether a household purchases broadband service. The model examined two types of factors: the tax status of states in which respondents live, and the characteristics of households. We also discussed these issues, as well as the influence of characteristics and uses of broadband service, with stakeholders.

Based on our model and interviews with stakeholders, we identified several characteristics of households that influence broadband adoption. First, our model indicated that high-income households are 39 percentage points more likely to purchase broadband service than are low-income households.[24] Similarly, some stakeholders we spoke with stated that adoption of broadband service is more widespread in communities with high income levels. A key underlying factor may be that computer ownership is substantially higher among higher-income households, according to a survey conducted by the Census Bureau. Second, our model results showed that households with a college graduate are 12 percentage points more likely to subscribe to broadband services compared with households without a college graduate. In fact, when discussing the effects of education on the demand for broadband, we were told that some college graduates see broadband as a necessity and would be less likely to choose to live in a rural area that did not have adequate broadband facilities. Third, we found that households headed by young adults are more likely to purchase broadband than are households headed by a person 50 or older.[25] Similarly, a few stakeholders we spoke with said that older adults are less likely to purchase broadband. This may be the case because older Americans generally have lower levels of computer ownership and computer familiarity. We also were told that households with children in school are more likely to have broadband service. Figure 4 provides some descriptive statistics to illustrate the relationship between several demographic characteristics and the adoption of broadband.

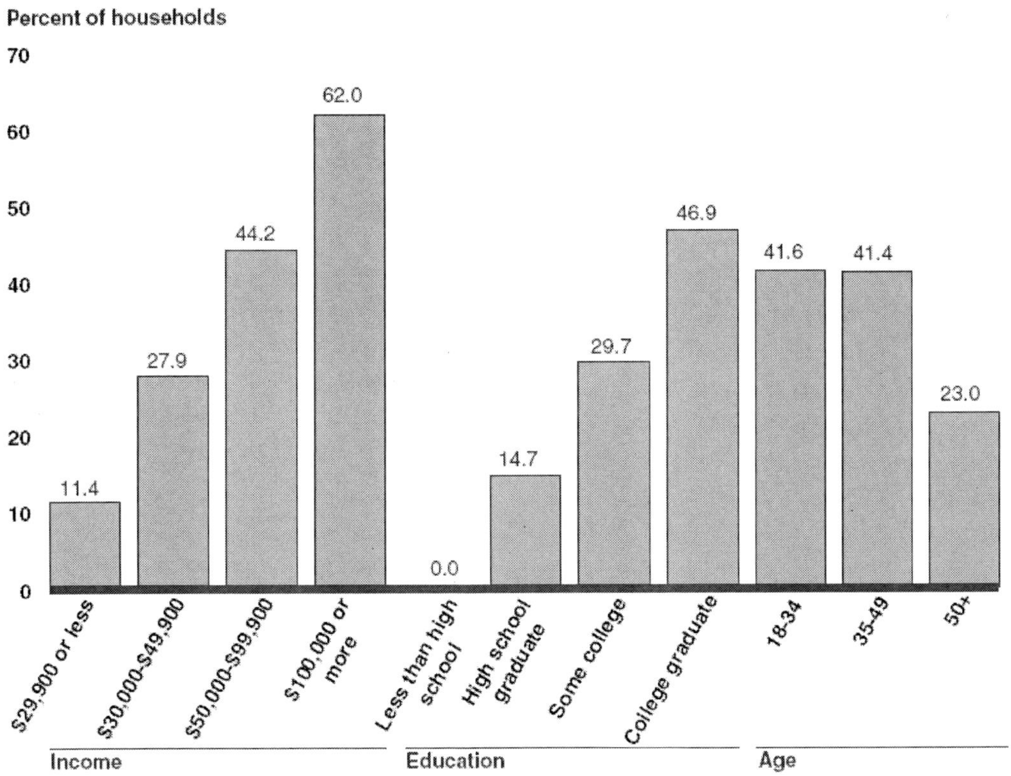

Source: GAO analysis of Knowledge Networks/SRI's The Home Technology MonitorTM: Spring 2005 Ownership and Trend Report.

Figure 4. Factors Influencing Subscription to Broadband.

We also examined whether households residing in rural areas were less likely to purchase broadband service than those living in urban areas. As noted earlier, we found that only 17 percent of rural households subscribe to broadband service. Our model indicated, however, that when the availability of broadband to households, as well as demographic characteristics, are taken into account, rural households no longer appear less likely than urban households to subscribe to broadband. That is, the difference in the subscribership to broadband among urban and rural households appears to be related to the difference in availability of the service across these areas, and not to a lower disposition of rural households to purchase the service.

In addition to household characteristics, we also found that characteristics and uses of broadband service available to consumers can also influence the extent to which households purchase broadband service.

- Some stakeholders we spoke with mentioned that the price of broadband service is an important factor affecting a household's decision to purchase this service. Some stakeholders mentioned, for example, that one of the key reasons for the recent surge in DSL subscribership is due to recent price declines for the service: Some providers are now offering DSL for less than $15 per month. Conversely, because satellite broadband service is expensive and also requires the upfront purchase of expensive equipment needed to receive the satellite signal, several of those we spoke with said that the expense of satellite broadband deters its purchase. In fact, a recent study suggests that areas served by multiple providers, where prices may tend to be lower, may have higher rates of broadband adoption.[26] However, because we lacked data on the price of broadband service, we were unable to include this variable in our econometric model.[27] We did not find that the number of companies providing broadband service affected the likelihood that a household would purchase broadband service.

- Some stakeholders also told us that the availability of applications and content not easily accessible through dial-up, as well as the degree to which consumers are aware of and value this availability, contribute to a household's decision to adopt broadband. For example, some functions, applications, and content—such as gaming, VoIP, and music and video downloads—either need or function much more effectively with broadband service than with dial-up service and, as such, make broadband a major attraction for households that value these types of services and content. Alternatively, some applications, such as e-mail, function adequately with dial-up service, and for households that primarily use the Internet for e-mail, there may be little need to upgrade to broadband service. Several of those we spoke with noted that a "killer application"—one that nearly everyone would view as essential and might entice more American households to adopt broadband—has not yet emerged.
- We also examined whether the tax status of the state in which each survey respondent lived influenced their likelihood to adopt broadband service. As mentioned earlier, we used a binary variable to represent the presence of Internet taxation. As such, the variable may capture the influence of other characteristics of the states in which the households resided, in addition to the influence of the tax. Further, lacking a variable for the price of broadband service, we cannot assess how the imposition of the tax influenced the price of the service. Using our model, we found that the parameter estimate had the expected sign— indicating that the imposition of the tax may have reduced the likelihood that a household would purchase broadband service. While the estimate was not statistically significant at the 5 percent level, it was statistically significant at the 10 percent level, perhaps suggesting that it is a weakly significant factor. However, given the nature of our model, it is unclear whether this finding is related to the tax or other characteristics of the states in which households resided.

STAKEHOLDERS IDENTIFIED SEVERAL OPTIONS TO ADDRESS THE LACK OF BROADBAND IN CERTAIN AREAS, BUT CHALLENGES EXIST WITH IMPLEMENTATION

Stakeholders we spoke with identified several options to facilitate greater broadband service in unserved areas; however, each option poses special challenges. RUS broadband programs provide a possible means for targeted assistance to unserved areas, but stakeholders raised concerns about the effectiveness of the loan program and its eligibility criteria. USF programs have indirectly facilitated broadband deployment in rural areas, but it is unclear whether the program should be expanded to directly support broadband service. Finally, wireless technologies could help overcome some of the cost and technological limitations to providing service in remote locations, but congestion and the management of the spectrum remain possible barriers.

RUS Broadband Programs Could Provide a Source of Targeted Assistance, but Stakeholders Identified Several Concerns with the Programs

As mentioned earlier, RUS provides support through grants and loans to improve rural infrastructures providing broadband service. The Community Connect Broadband grant program provides funding for communities where no broadband service currently exists. One loan program, which provides loans at 4 percent, also requires that no existing broadband providers be present in a community, but loans at the Treasury interest rate are available to entities that plan to serve communities with existing broadband service. Several stakeholders with whom we spoke, as well as the findings of a recent report by the Inspector General (IG) of the Department of Agriculture, raised concerns about these programs:[28]

Effectiveness of Loans

It is not clear whether a loan program—such as the RUS loan program—is effective for helping rural areas gain access to broadband services. RUS requires applicants to submit an economically viable business plan—that is, applicants must show that their business will be sufficiently successful such that the applicant will be capable of repaying the loan. But developing a viable broadband business plan can be difficult in rural areas, which have a limited number of potential subscribers. As a result, RUS has rejected many applications because the applicant could not show that the business plan demonstrated a commercially viable and sustainable business. In fact, the agency has been unable to spend all of its loan program funds. Since the inception of the program in 2002, the agency has fallen far short of obligating the available funding in this program. For example, RUS officials told us that in 2004, they estimated that the appropriations for the broadband loan program could support approximately $2.1 billion in loans, but only 28 percent of this amount—or $603 million—was awarded for broadband projects. RUS officials also told us that its 2005 appropriations could support just over $2 billion in loans, but only 5 percent—or $112 million—was awarded to broadband projects. One stakeholder we spoke with suggested that a greater portion of RUS funds should be shifted from loans to grants in order to provide a more significant level of assistance for rural broadband deployment. RUS officials noted that they are currently evaluating the program and recognize that the program criteria limit the ability of the agency to utilize their full loan funding.

Competitive Environment Requirements

During our interviews, some stakeholders expressed concerns about how the presence of existing broadband deployment was considered in evaluating RUS grant and loan applications. In the case of the grant program, RUS approves applications only for communities that have no existing broadband service. Some local government officials and a company we spoke with noted that this "unserved" requirement for RUS grants can disqualify certain rural communities that have very limited Internet access—perhaps in only one small part of a community.[29] Alternatively, regarding the Treasury rate loan program, a few providers and the IG's report criticized the program for supporting the building of new infrastructure where infrastructure already existed. In particular, we learned that loans were being let for deployment in areas that already had at least one provider and in some cases had several providers. As such, it is not clear whether these funds are being provided to communities most in need. RUS officials noted, however, that the statute specifically allows such loans. Additionally, the issue of how the status of existing service is gauged was a concern for one provider we spoke with. RUS obtains information about existing providers from applicants, and agency officials told us that agency field representatives review the veracity of information provided by applicants during field visits. However, RUS officials told us that FCC zip-code data is not granular enough for their needs in evaluating the extent of broadband deployment in rural areas.

Community Eligibility

A few local officials we spoke with criticized the community size and income eligibility requirements for the grant and loan programs. In Massachusetts, one stakeholder said that most small towns in part of that state exceed RUS's population requirements and thus do not qualify for grants or loans. The grant and loan programs also have per-capita personal income requirements. One service provider in Alaska said that the grant program income eligibility requirements can exclude Alaskan communities, while failing to take into account the high cost of living in rural Alaska.

Technological Neutrality

Satellite companies we spoke with said RUS's broadband loan program requirements are not readily compatible with their business model or technology. Once a company launches a satellite, the equipment that individual consumers must purchase is the remaining infrastructure expense. Because the agency requires collateral for loans, the program is more suited for situations where the *providers*, rather than individual consumers, own the equipment being purchased through the loan. Yet, when consumers purchase satellite broadband, it is common for them to purchase the equipment needed to receive the satellite signal, such as the reception dish. Additionally, broadband service must be provided at a speed of at least 200 kilobits *in both directions*—which is not necessarily the case for satellite broadband—for it to qualify for RUS loans. Moreover, RUS officials noted that for satellite broadband providers to be able to access RUS loans, they would have to demonstrate that each customer lives in a community that meets the community size eligibility requirement. As such, this program may not be easily utilized by satellite broadband providers. Yet for some places, satellite could be a cost-effective mechanism to provide broadband infrastructure into rural areas. For example, in 2005, the RUS Community Connect program provided grants to 19 communities that average 554 residents and 194 households. The total cost of these grants was roughly $9 million. Thus, RUS spent an average of $2,443 per covered household,[30] but the cost per household that adopted broadband would be even higher since only a subset of these households would choose to subscribe to broadband service. By contrast, two satellite providers we spoke with estimate that their consumer equipment and installation costs are roughly $600 per subscribing household. These figures might not fully represent the full nature of the services provided through the grant program and those available via satellite; for example, grantees of the RUS program are required to provide free Internet service to community centers.

USF Programs Indirectly Support Broadband Service, but Several Stakeholders Expressed Concerns

While the USF program does not directly fund broadband service, the funding provided to support telecommunications networks indirectly supports the development of infrastructure that can provide many communications services, including broadband. USF's high-cost program helps maintain and upgrade telecommunications networks in rural areas. Three stakeholders we spoke with in Alaska, Ohio, and North Dakota attributed the relative success of broadband deployment in rural areas to the USF program. Additionally, the Schools and Libraries Program and the Rural Health Care Program help facilitate broadband service to specific locations; according to two providers in Alaska, these programs have been very beneficial in bringing some form of broadband service to rural Alaskan villages that might have received no service without these government programs.

However, stakeholders we spoke with identified several concerns about the USF program:

- Large ILECs serving rural areas and rural ILECs receive high-cost fund support under different formulas. The two types of ILECs have different eligibility criteria under which they can qualify to receive high-cost support and more support is provided to rural companies than to nonrural companies serving rural areas.[31] Two stakeholders we spoke with suggested that the eligibility criteria should be modified, such that the criteria better reflect the cost to provide service in particular areas, rather than the type of company providing the service. Alternatively, two stakeholders we spoke with favor the current eligibility criteria and funding mechanism.
- Two stakeholders we spoke with expressed concerns about a lack of coordination across USF funding sources, which could lead to inefficient use of funds and inadequate leveraging of funds. For example, in Alaska, two stakeholders noted that governments and providers receive "silos" of funding for schools, libraries, and rural health centers. Because the programs are narrowly defined, multiple entities might be

the recipient of funding for broadband service, which could lead to multiple broadband connections in relatively small rural communities. One stakeholder noted that since each entity might use only a fraction of its available broadband capacity, there can be capacity for Internet traffic available for other uses or users, but funding recipients are sometimes not allowed to share this capacity, either with other entities or with residents in the community. Thus, communities may be unable to leverage the available funding for other uses.

- While two stakeholders we spoke with suggested expanding the USF program to include broadband service, we found little support for this overall. Some stakeholders we spoke with expressed concern about funding the USF program at current levels of support. These stakeholders fear that expanding the USF program to include broadband service, which would increase program expenditures and thus require additional funding, could undermine support for the entire USF program.

Resolving Spectrum Congestion and Management Concerns Could Facilitate Greater Wireless Broadband Service

As mentioned previously, certain wireless technologies hold the potential for supporting broadband service in difficult-to-serve rural areas. In less densely populated areas, installing wire-based facilities for cable modem and DSL service represents a significant cost factor. Therefore, certain wireless technologies may be a lower-cost way to serve rural areas than wireline technologies.

While wireless technologies hold the promise of expanding the availability of broadband, some stakeholders we spoke with expressed concern about the degree of congestion in certain bands as well as the management of spectrum. For example, in some geographic areas, we heard that congestion in certain unlicensed spectrum bands makes providing wireless broadband Internet access more difficult, and a few stakeholders said that with more unlicensed spectrum, wireless providers could support greater broadband deployment. Additionally, wireless providers we spoke with also expressed concern about the management of spectrum, particularly the quality of certain bands and quantity of spectrum available for wireless broadband service. Two stakeholders mentioned that spectrum allocated to wireless broadband service is susceptible to having communications obstructed by interference from trees and buildings. In a 2005 report, we noted that experts agreed that the government should evaluate its allocation of spectrum between licensed and unlicensed uses.[32] But we also noted that these experts failed to agree on whether FCC should dedicate more or less spectrum to unlicensed uses. In June 2006, FCC will conduct an auction of spectrum dedicated to advanced wireless services, which will make available 90 MHz of spectrum for wireless broadband services. FCC staff also noted that the commission has other efforts underway to increase available spectrum for wireless broadband services.

CONCLUSION

In the past several years, the importance of broadband for Americans and for the American economy has been articulated by interested stakeholders, as well as by the President, Congress, and the last several FCC chairmen. Universal availability of broadband has been set forth as a policy goal for the near term—2007. And progress toward this goal has been substantial. The availability of broadband to residential consumers has grown from its nascent beginnings in the latter part of the 1990s to broad coverage throughout the country. In the last 10 years, providers in traditional communications industry segments—telephone and cable—have upgraded and redesigned miles of their networks in order to offer broadband services. The provision of broadband through various wireless means, as well as over the existing electricity infrastructure, have also been developed, and for many, if not most Americans, the burgeoning broadband marketplace is characterized by competitive choice in broadband access and creative and ever-expanding applications and content. Many would consider the rollout of broadband infrastructure as a success story of entrepreneurial initiative.

But not all places or people have experienced the full benefits of this rapid rollout of broadband services. As with many other technologies, the costs of bringing broadband infrastructure to rural America can be high. For private providers who must weigh the costs and returns of their investments, the feasibility of serving the most rural parts of our country may not work within a reasonable business model. While there are federal support mechanisms for rural broadband, it is not clear how much impact these programs are having or whether their design suggests a broad consideration of the most effective means of addressing the problem. And one of the difficulties of assessing the gaps in deployment and where to target any federal support is that it is hard to know exactly where broadband infrastructure has not been deployed. FCC does collect data on the geographic extent of providers' service, but these data are not structured in a way that accurately illustrates the extent of deployment to residential users. Without accurate, reliable data to aid in analysis of the existing deployment gaps, it will be difficult to develop policy responses toward gaps in broadband availability. This could hinder our country's attainment of universally available broadband. And as the industry moves quickly to even higher bandwidth broadband technologies, we risk leaving some of the most rural places in America behind.

RECOMMENDATION FOR EXECUTIVE ACTION

In a draft of this report provided to FCC for review and comment, GAO recommended that FCC identify and evaluate strategies for improving the 477 data such that the data provide a more accurate depiction of residential broadband deployment throughout the country. In oral comments regarding this recommendation, FCC staff acknowledged that the 477 data have some limitations in detailing broadband deployment, but also noted that there had recently been a proceeding examining its broadband data collection efforts and that some changes to the data collection had been implemented. In that proceeding, the commission also determined that it would be costly and could impose large burdens on filers—particularly small entities—to require any more detailed filings on broadband deployment. Although FCC staff told us that analysis of potential costs had been conducted, exact estimates of these costs and burdens have not yet been determined. Moreover, many have expressed concern about ensuring that all Americans—especially those in rural areas—have access to broadband technologies. Policymakers concerned about full deployment of broadband throughout the country will have difficulty targeting any assistance to that end without accurate and reliable data on localized deployment. As such, we recommend that FCC develop information regarding the degree of cost and burden that would be associated with various options for improving the information available on broadband deployment and should provide that information to the Senate Committee on Commerce, Science, and Transportation and the House Energy and Commerce Committee in order to help them determine what actions, if any, are necessary to employ going forward.

Agency Comments and our Evaluation

We provided a draft of this report to the Department of Agriculture, the Department of Commerce, and the Federal Communications Commission for their review and comment. The Department of Agriculture provided no comments. The Department of Commerce and FCC provided technical comments that we incorporated, as appropriate. FCC did not comment on the final recommendation contained in this report.

We also provided a draft of this report to several associations representing industry trade groups and state and local government entities for their review and comment. Specifically, the following associations came to GAO headquarters to review the draft: Cellular Telecommunications and Internet Association (CTIA), National Association of Regulatory Utility Commissioners (NARUC), National Association of Telecommunications Officers and Advisors (NATOA), National Cable and Telecommunications Association (NCTA), National Telecommunications Cooperative Association (NTCA), Satellite Industry Association (SIA), US Internet Industry Association (USIIA), United States Telecom Association (USTA), and Wireless Internet Service Providers Association (WISPA). Officials from CTIA, NARUC, and NTCA did not provide comments. Officials from NATOA, NCTA, SIA, and USIIA provided technical comments that were incorporated, as appropriate. USTA and WISPA provided comments that are discussed in appendix V.

We are sending copies of this report to the appropriate congressional committees and to the Secretary of Agriculture, the Secretary of Commerce, and the Chairman of the Federal Communications Commission. We will also make copies available to others upon request. In addition, the report will be available at no charge on the GAO Web site at http://www.gao.gov.

If you have any questions about this report, please contact me at (202) 512- 2834 or heckerj@gao.gov. Contact points for our Offices of Congressional Relations and Public Affairs may be found on the last page of this report. Contact information and major contributors to this report are listed in appendix VI.

JayEtta Z. Hecker
Director, Physical Infrastructure Issues

APPENDIX I. SCOPE AND METHODOLOGY

The objectives of the report were to provide information on (1) the current status of broadband deployment and adoption, (2) the factors that influence the deployment of broadband networks, (3) the factors that influence the adoption of broadband service by households, and (4) the options that have been suggested to spur greater broadband deployment and adoption. To respond to the four objectives, we used a variety of approaches.

To gather opinions for all four objectives, we employed a case-study approach. This approach allowed us to identify issues at the state and local level that would not be apparent in nationwide data. We selected eight states for our case studies: Alaska, California, Kentucky, Massachusetts, North Dakota, Ohio, Texas, and Virginia. We selected these states based on Census Bureau data on statewide income, urbanization, population density, and percentage of households using the Internet. We also considered whether each state taxed Internet access. We sought to include states in diverse categories of each of our selection criteria. In each state, we interviewed state and local officials, including local franchising authorities, state public utility regulators, representatives from state governor's offices; associations; private cable and telephone providers; wireless Internet service providers; and municipal and cooperative telecommunications providers.

We also spoke with a variety of individuals and organizations knowledgeable about broadband services. In particular, we spoke with industry providers, trade associations, and academic experts. We also spoke with representatives from the Federal Communications Commission (FCC), the National Telecommunications and Information Administration of the Department of Commerce, and the Rural Utilities Service of the Department of Agriculture.

To assess the factors influencing the deployment and adoption of broadband, we used survey data from Knowledge Networks/SRI's *The Home Technology Monitor™: Spring 2005 Ownership and Trend Report*. Knowledge Networks/SRI is a survey research firm that conducted a survey on household ownership and use of consumer electronics and media. Knowledge Networks/SRI interviewed approximately 1,500 randomly sampled telephone households, asking questions about the household's purchase of computers and Internet access. All percentage estimates from the Knowledge Networks/SRI survey have margins of error of plus or minus 7 percentage points or less, unless otherwise noted. See appendix II for a discussion of the steps we took to evaluate the reliability of Knowledge Networks/SRI's data. Using the data from Knowledge Networks/SRI, we estimated two econometric models. One model examined the factors affecting broadband deployment. We also developed a model to examine the factors affecting a household's adoption of broadband services. See appendix III for a more detailed explanation of, and results from, our deployment and adoption models.

To assess the status of broadband deployment, we used FCC's Form 477 data that identified companies providing broadband service by zip code. We used FCC's data to identify the companies reporting to provide broadband service in the zip codes where respondents to Knowledge Networks/SRI's survey resided. To assess the reliability of FCC's Form 477 data, we reviewed documentation, interviewed knowledgeable officials, and performed electronic testing of the data elements used in our analyses. We made several adjustments to these data, such as excluding satellite companies and companies only providing service to businesses. See appendix III for more on our methodology concerning adjustment to FCC's 477 data. With these adjustments to the data, we determined that they were sufficiently reliable for the purposes of this report.

We conducted our work from April 2005 through February 2006 in accordance with generally accepted government auditing standards.

APPENDIX II. DATA RELIABILITY

To obtain information on the types of Internet access purchased, or adopted, by U.S. households, we purchased existing survey data from Knowledge Networks Statistical Research (Knowledge Networks/SRI). Their survey was completed with 1,501 of the estimated 3,127 eligible sampled households for a response rate of 48 percent. The survey was conducted between February 22 and April 15, 2005.

The study procedures yielded a sample of members of telephone households in the continental United States using a national random-digit dialing method. Survey Sampling Inc. (SSI) provided the sample of telephone numbers, which included both listed and unlisted numbers and excluded blocks of telephone numbers determined to be nonworking or business-only. At least five calls were made to each telephone number in the sample to attempt to interview a responsible person in the household. Special attempts were made to contact refusals and convert them into interviews; refusals were sent a letter explaining the purpose of the study and an incentive. Data were obtained from telephone households and are weighted to the total number of households in the 2005 Current Population Survey adjusted for multiple phone lines.

As with all sample surveys, this survey is subject to both sampling and nonsampling errors. The effect of sampling errors due to the selection of a sample from a larger population can be expressed as a confidence interval based on statistical theory. The effects of nonsampling errors, such as nonresponse and errors in measurement, may be of greater or lesser significance but cannot be quantified on the basis of available data.

Sampling errors arise because of the use of a sample of individuals to draw conclusions about a much larger population. The study's sample of telephone numbers is based on a probability selection procedure. As a result, the sample was only one of a large number of samples that might have been drawn from the total telephone exchanges from throughout the country. If a different sample had been taken, the results might have been different. To recognize the possibility that other samples might have yielded other results, we express our confidence in the precision of our particular sample's results as a 95 percent confidence interval. We are 95 percent confident that when only sampling errors are considered each of the confidence intervals in this report will include the true values in the study population. All percentage estimates from the survey have margins of error of plus or minus 7 percentage points or less, unless otherwise noted. The 95 percent confidence interval for the estimate of the total number of U.S. households that subscribed to broadband service in 2005 is 28.5 million to 33.7 million households.

In addition to the reported sampling errors, the practical difficulties of conducting any survey introduce other types of errors, commonly referred to as nonsampling errors. For example, questions may be misinterpreted, some types of people may be more likely to be excluded from the study, errors could be made in recording the questionnaire responses into the computer-assisted telephone interview software, and the respondents' answers may differ from those who did not respond. Knowledge Networks/SRI has been fielding versions of this survey for over 20 years. In addition, to reduce measurement error, Knowledge Networks/SRI employs interviewer training, supervision, and monitoring, as well as computer- assisted interviewing to reduce error in following skip patterns.

For this survey, the 48 percent response rate is a potential source of nonsampling error; we do not know if the respondents' answers are different from the 52 percent who did not respond. Knowledge Networks/SRI took steps to maximize the response rate—the questionnaire was carefully designed and tested through deployments over many years, at least five telephone calls were made at varied time periods to try to contact each telephone number, the interview period extended over about 8 weeks, and attempts were made to contact refusals and convert them into interviews.

Because we did not have information on those contacted who chose not to participate in the survey, we could not estimate the impact of the nonresponse on our results. Our findings will be biased to the extent that the people at the 52 percent of the telephone numbers that did not yield an interview have different experiences with Internet access than did the 48 percent of our sample who responded. However, distributions of selected household characteristics (including presence of children, race, and household income) for the sample and the U.S. Census estimate of households show a similar pattern.

To assess the reliability of these survey data, we relied on a prior GAO report that made use of the Knowledge Networks/SRI 2004 survey for a similar purpose. In that prior assessment, we determined that the data were sufficiently reliable for our purposes. For this report we reviewed Knowledge Networks/SRI's documentation of survey procedures for 2005 and compared them to the procedures used in their 2004 survey. We determined that their survey methodology was substantively unchanged.

Additionally, we performed electronic testing of the 2005 survey data elements used in this report. We determined that the data were sufficiently reliable for the purposes of this report.

Appendix III. Broadband Deployment and Adoption Models

This appendix describes our models of broadband deployment and adoption. Specifically, we discuss (1) the design of our models, (2) the data sources, (3) our methodology for assessing broadband deployment, and (4) the estimation methodology and results.

Design of our Broadband Deployment and Adoption Models

A company will deploy broadband service in an area only if the company believes that such a deployment will be profitable. Similarly, a household will purchase, or adopt, broadband service only if the value, or utility, to members of the household exceeds the price the household must pay to receive the service. In this section, we explain the two models we developed to examine the factors that influence the deployment and adoption of broadband service.

Deployment Model

A company will deploy broadband service in an area only if the company believes that such a deployment will be profitable. Based on conversations with industry stakeholders, including companies deploying broadband service, we identified a number of factors that influence a company's decision to deploy broadband service. In particular, the following factors may influence the decision to deploy broadband service: population density, terrain, backhaul costs, existing or potential competition, the technical expertise of the population, the income of the population, and regulatory policies (such as rights-of-way policies). We also reviewed relevant studies, and noted the same and additional factors that may influence the deployment of broadband service.[1] Some of these factors, such as the population density and backhaul, will influence the cost of providing broadband service, while other factors, such as the income of the population, will influence the potential revenues that a company may hope to generate. Together, these revenue and cost factors will influence the potential profitability of providing broadband service, and ultimately the decision to deploy broadband service.

To empirically test these hypotheses, we estimated the following econometric model; since all the variables identified above were not available, we were unable to include some of the variables—such as terrain—in our model. The decision to deploy broadband service is a function of

- the population in the area;
- the population density in the area;
- the percentage of the population residing in an urban area;
- the per-capita income in the area;
- the educational attainment of the population in the area;
- the population teleworking in the area;
- the age of the population in the area;
- the distance to a metropolitan area with a population of 250,000 or more; and
- whether the state in which the area is located imposed a tax on Internet access in 2005.

Adoption Model

Households will purchase, or adopt, broadband service only if the value, or utility, that members of the household receive from the service exceeds the price of the service. In conversations with industry stakeholders, we were told that several characteristics of households influence the extent to which households purchase broadband service; we also reviewed other studies, and noted characteristics of households that these studies associated with the purchase of broadband service.[2] In particular, the following characteristics of

[1] For example, see Tony H. Grubesic and Alan T. Murray, "Waiting for Broadband: Local Competition and the Spatial Distribution of Advanced Telecommunication Services in the United States," *Growth and Change*, vol. 35, no. 2 (2004): 139-165; and James E. Prieger, "The Supply Side of the Digital Divide: Is There Equal Availability in the Broadband Internet Access Market?" *Economic Inquiry*, vol. 41, no. 2 (2003): 346-363.

[2] For example, see Scott Wallsten, *Broadband Penetration: An Empirical Analysis of State and Federal Policies* (Washington, D.C.: AEI-Brookings Joint Center for Regulatory Studies, 2005); Scott J. Savage and Donald M.

households may influence the decision to purchase broadband service: income, education, age of household members, presence of children in the household, and the technological knowledge of members of the household. These characteristics may be associated with the extent to which a household would benefit from, and therefore value, broadband service, such as using broadband to telework, conduct research for school, and playing games. Industry stakeholders also noted that price influences a household's decision to purchase broadband service.

To empirically test these hypotheses, we estimated the following econometric model; because we lacked data on the price of broadband service, we were unable to include this variable in our econometric model.[3] The decision to purchase, or adopt, broadband service is a function of

- the income of the household;
- the education attainment of the heads of the household;
- the age of the heads of the household;
- the presence of children in the household;
- the racial composition of the household;
- the occupation of the heads of the household;
- the number of people in the household;
- whether the household resides in an urban, suburban, or rural location;
- the number of companies providing broadband service in the area; and
- whether the state in which the household resides imposes a tax on Internet access.

Data Sources

We required several data elements to build the data set used to estimate our deployment and adoption models. The following is a list of our primary data sources. In addition, we list all of the variables, definitions, and sources for the deployment model in table 1 and the adoption model in Table 2.

- We obtained data on a sample of households in the United States from Knowledge Networks/SRI, using Knowledge Networks/SRI's product *The Home Technology Monitor™: Spring 2005 Ownership and Trend Report.* From February through April 2005, Knowledge Networks/SRI interviewed a random sample of 1,501 households in the United States. Knowledge Networks/SRI asked participating households a variety of questions about their use of technology, including questions such as whether the household purchased broadband service, and about the household's demographic characteristics.

Waldman, "United States Demand for Internet Access," *Review of Network Economics*, vol. 3, no. 3 (2004): 228-247; and Debra J. Aron and David E. Burnstein, "Broadband Adoption in the United States: An Empirical Analysis" (paper presented at the 31st Annual Telecommunications Policy Research Conference, Arlington, Va., 2003).

[3] As such, we recognize that the parameter estimates will be biased. We are unable to assess the possible extent of this bias.

- From the Federal Communications Commission (FCC), we obtained information on the companies providing broadband service in zip codes throughout the United States in December 2004. For each zip code, FCC provided the names of companies reporting, through the agency's Form 477, that they provided broadband service to at least one residential or small business customer and the type of company providing the service (e.g., cable and satellite).
- We used the most recent information from the U.S. Census Bureau to obtain demographic information for the areas where the households responding to Knowledge Networks/SRI's survey were located.

Table 1. Deployment Model: Definitions and Sources of Variables

Variable	Definition	Source
Deploy	A binary variable that equals 1 if broadband service is available to the household responding to Knowledge Networks/SRI's survey.	FCC 2004 Form 477and GAO analysis
Internet taxation	A binary variable that equals 1 if the state where the household resides imposes a tax on Internet access.	GAO analysis
Population, in thousands	The number of residents in the area where the household resides, in thousands.	Census Bureau
Urbanization	The percentage of the population residing in an urban area.	Census Bureau
Distance	The distance to a metropolitan area with a population of 250,000 or more.	GAO analysis
Percentage of work-at-home residents	The percentage of the population working from home.	Census Bureau
Percentage of population under 16	The percentage of the population under the age of 16.	Census Bureau
Percentage of population 17 to 24	The percentage of the population 17 to 24 years old.	Census Bureau
Percentage of population 65 or older	The percentage of the population 65 or older.	Census Bureau
Percentage of population with a high-school degree	The percentage of the population with a high-school degree.	Census Bureau
Percentage of population with education beyond high school	The percentage of the population with education beyond high school.	Census Bureau
Per-capita income, in thousands	The per-capita income in the area, in thousands of dollars.	Census Bureau
Population density, in thousands	The ratio of population to square miles in the area, in thousands.	Census Bureau

Source: GAO.

Table 2. Adoption Model: Definitions and Sources of Variables

Variable	Definition	Source
Adopt	A binary variable that equals 1 if the household responding to Knowledge Networks/SRI's survey purchases broadband service.	Knowledge Networks/SRI
Internet taxation	A binary variable that equals 1 if the state where the household resides imposes a tax on Internet access.	GAO analysis
Number of broadband providers	The number of companies providing broadband service to the household.	FCC 2004 Form 477 and GAO analysis
Income between $30,000 and $49,900	A binary variable that equals 1 if the household's income is between $30,000 and $49,900.	Knowledge Networks/SRI
Income between $50,000 and $99,900	A binary variable that equals 1 if the household's income is between $50,000 and $99,900.	Knowledge Networks/SRI
Income $100,000 or more	A binary variable that equals 1 if the household's income is greater than or equal to $100,000.	Knowledge Networks/SRI
Race-white	A binary variable that equals 1 if the household reported its race as white.	Knowledge Networks/SRI
College graduate	A binary variable that equals 1 if either the man or woman of the household is a college graduate.	Knowledge Networks/SRI
Age 34 to 49	A binary variable that equals 1 if either the man or woman of the household is 34 to 49 years old, and neither is younger than 34.	Knowledge Networks/SRI
Age 50 or older	A binary variable that equals 1 if both the man and woman of the household are 50 years old or older.	Knowledge Networks/SRI
Children	A binary variable that equals 1 if a child age 17 or younger resides in the home.	Knowledge Networks/SRI
Household size	A binary variable that equals 1 if three or more people reside in the home.	Knowledge Networks/SRI
Occupation-professional	A binary variable that equals 1 if the man or woman of the household reports working in a professional position.	Knowledge Networks/SRI
Occupation-clerical, sales, or technical	A binary variable that equals 1 if the man or woman of the household reports working in a clerical, sales, or technical position, and neither reports working in a professional position.	Knowledge Networks/SRI
Occupation-blue collar	A binary variable that equals 1 if the man and woman of the household report working in a blue collar position.	Knowledge Networks/SRI

Table 2. Continued

Variable	Definition	Source
Occupation-other	A binary variable that equals 1 if the man or woman of the household reports working in a position other than a professional, clerical, sales, technical, or blue-collar position.	Knowledge Networks/SRI
Rural location	A binary variable that equals 1 if the household resides in an area outside a metropolitan statistical area (MSA).	GAO analysis
Suburban location	A binary variable that equals 1 if the household resides in an area inside an MSA but outside the central city of that MSA.	GAO analysis

ASSESSING BROADBAND

Deployment FCC's Form 477 data include information on companies providing broadband service to at least one residential or business customer in zip codes throughout the United States in December 2004. However, since zip codes can represent large geographic areas, companies providing broadband service in a zip code might not have facilities in place to serve all households in the zip code. Thus, while a household might reside in a zip code in which FCC's Form 477 indicates that broadband service is available, that service might not be available to the household. Additionally, as we note in the text, we identified other concerns with FCC's data. Therefore, we took additional steps to assess whether broadband service was available to households included in Knowledge Networks/SRI's survey. In particular, we took the following steps for each observation in our data set:

- removed firms providing only satellite service;
- removed firms that provided only broadband service to business customers, since residential households were the focus of our study;
- removed large incumbent local exchange carriers when the company was identified as providing service in areas that lay outside of its local
- exchange area, since these firms typically provide service only to business customers outside of their local exchange areas;[4]
- removed firms when 2 or more of the 10 largest cable operators reported providing broadband service, since large cable operators rarely have overlapping service territories;
- removed cable operators if the responding household indicated that cable service did not pass the residence; and
- removed companies providing telephone-based broadband service if the household's residence was greater than 2.5 miles from the central office facility, since DSL service is distance limited.

---

[4] We did not remove Verizon, since thought its acquisition of GTE, it serves a wide variety of locations as an incumbent exchange carrier.

ESTIMATION METHODOLOGY AND RESULTS

For both the deployment model and adoption model, we are estimating a reduced-form, binary-choice model. That is, broadband service is either deployed in the area or it is not, and the household either purchases broadband service or it does not. Given the binary choice nature of the models, we employed the probit method to estimate the deployment and adoption equations.[5] In this section, we present descriptive statistics and estimation results for the two equations and discuss the results.

Deployment Model

In table 3, we provide basic statistical information on all of the variables included in the deployment model, and in table 4, we provide the results from the probit estimation of the deployment model. Of the 1,501 respondents to Knowledge Networks/SRI's survey, we used 1,402 observations in the deployment model; we were unable to match the zip +4 code for all 1,501 observations with publicly available data, which was necessary to assess whether the residence was 2.5 miles from the serving central office facility.

Table 3. Deployment Model: Descriptive StatisticsSource:

Variable	Mean	Standard deviation	Minimum value	Maximum value
Deploy	0.911	0.285	0.000	1.000
Internet taxation	0.546	0.498	0.000	1.000
Population, in thousands	26.022	17.982	0.070	113.935
Urbanization	76.154	33.240	0.000	100.000
Distance	34.361	42.743	0.249	572.803
Percentage of work-at-home residents	3.257	2.064	0.000	33.333
Percentage of population under 16	24.018	4.651	6.225	41.219
Percentage of population 17 to 24	10.585	4.785	0.583	55.113
Percentage of population 65 or older	12.995	5.688	2.195	59.057
Percentage of population with a high-school degree	29.546	9.002	3.121	68.966
Percentage of population with education beyond high school	51.395	16.092	8.836	95.348
Per-capita income, in thousands	44.466	15.597	9.583	164.479
Population density, in thousands	2.976	6.876	0.002	74.814

Source: GAO.

[5] An alternative method to estimate these equations is the logit model. In a binary choice model, the differences between the logit and probit models are generally not significant. Differences can arise in the multinomial model, where there are three or more choices, because the logit model imposes independence conditions that sometimes do not reflect the conditions being modeled. Such was not the case in our models, since we are estimating binary choice equations.

Table 4. Deployment Model: Estimation Results

Variable	Parameter estimateand [p-value]
Intercept	-2.9299[0.0097]a
Internet taxation	-0.1486[0.2275]
Population, in thousands	0.0099[0.1140]
Urbanization	0.0102[0.0001]a
Distance	-0.0012[0.3115]
Percentage of work-at-home residents	-0.0600[0.0392]b
Percentage of population under 16	0.0335[0.1192]
Percentage of population 17 to 24	0.0198[0.3027]
Percentage of population 65 or older	0.0468[0.0271]b
Percentage of population with a high-school degree	0.0114[0.3260]
Percentage of population with education beyond high school	0.0121[0.1957]
Per-capita income, in thousands	0.0270[0.0074]a
Population density, in thousands	0.1706[0.0159]b
Number of observations	1,402
1-LogL/Log0	32.0077

Source: GAO.

[a] Significant at the 1 percent level.

[b] Significant at the 5 percent level.

Results from our model indicate that several factors related to the cost of providing broadband service and the demand for broadband service influence the likelihood that service will be available in a particular area. Regarding the cost factors, we found that urban areas and areas with greater population density are more likely to receive broadband service. For example, urban areas are about 9 percentage points more likely to receive broadband service than are similar rural areas. These results are consistent with broadband service being less costly to deploy in densely populated, more urban environments, where a similar investment in facilities can serve a greater number of subscribers than is possible in rural areas. Regarding demand for broadband service, we found that areas with greater per-capita incomes are more likely to receive broadband service. Additionally, we found that areas with a greater number of people working from home are less likely to have broadband service and that areas with a greater percentage of people age 65 or older are more likely to have broadband service.

We did not find that taxation of Internet access by state governments influenced the deployment of broadband service. Taxes can raise consumer prices and reduce revenues and impose costs on providers, and thereby possibly reduce the incentive for companies to deliver a product or service. Since we used a binary variable to indicate the presence of taxes, this variable could also potentially capture the influence of other characteristics of the states, in addition to the influence of the tax. Results from our model indicate that Internet access taxes do not affect the likelihood that companies will deploy broadband service; while the parameter estimate has the expected sign, the estimate is not statistically significant.

Table 5. Adoption Model: Descriptive Statistics

Variable	Mean	Standard deviation	Minimum value	Maximum value
Adopt	0.336	0.473	0.000	1.000
Internet taxation	0.553	0.497	0.000	1.000
Number of broadband providers	3.307	2.161	1.000	9.000
Income between $30,000 and $49,900	0.223	0.417	0.000	1.000
Income between $50,000 and $99,900	0.336	0.473	0.000	1.000
Income $100,000 or more	0.149	0.356	0.000	1.000
Race-white	0.858	0.349	0.000	1.000
College graduate	0.499	0.500	0.000	1.000
Age 34 to 49	0.378	0.485	0.000	1.000
Age 50 or older	0.424	0.494	0.000	1.000
Children	0.387	0.487	0.000	1.000
Household size	0.465	0.499	0.000	1.000
Occupation-professional	0.442	0.497	0.000	1.000
Occupation-clerical, sales, or technical	0.154	0.361	0.000	1.000
Occupation-blue collar	0.029	0.167	0.000	1.000
Occupation-other	0.244	0.430	0.000	1.000
Rural location	0.052	0.222	0.000	1.000
Suburban location	0.568	0.496	0.000	1.000

Adoption Model

In table 5, we provide basic statistical information on all of the variables included in the adoption model, and in table 6, we provide the results from the probit estimation of the adoption model. Since households can only chose to purchase, or adopt, broadband service where it is deployed, we only include households from Knowledge Networks/SRI's survey where we assessed that broadband service was available; based on our analysis, 133 respondents did not have broadband service available. Further, 355 respondents to Knowledge Networks/SRI's survey did not answer one or more demographic questions and 29 did not answer, or did not know, what type of Internet connection their household purchased. Therefore, we excluded these respondents. Thus, we used 901 observations in the adoption model.[6]

Our model results indicate that four characteristics influence whether households purchase, or adopt, broadband service. First, we found that households with greater incomes are more likely to purchase broadband service than are lower-income households. For example, the 25 percent of households with the highest income levels were about 39 percentage points more likely to purchase broadband service than the 25 percent of households with the lowest income levels. Second, households with a college graduate are

[6] We did not find that the households that failed to answer the demographic questions were more or less likely to be online than were the households that answered these questions.

about 12 percentage points more likely to purchase broadband service than are households without a college graduate. We also found that white households are more likely to purchase broadband service than households of other races. Finally, older households are less likely to purchase broadband service than are younger households.

As with the deployment model, we did not find that taxation of Internet access by state governments influenced the adoption of broadband service. As mentioned earlier, we used a binary variable to represent the presence of Internet taxation. As such, the variable may capture the influence of other characteristics of the states in which the households resided, in addition to the influence of the tax. Further, lacking a variable for the price of broadband service, we cannot assess how the imposition of the tax influenced the price of the service and thus the household's adoption decision. Using our model, we found that the parameter estimate had the expected sign—indicating that the imposition of the tax may have reduced the likelihood that a household would purchase broadband service. While the estimate was not statistically significant at the 5 percent level, it was statistically significant at the 10 percent level, perhaps suggesting that it is a weakly significant factor. However, given the nature of our model, it is unclear whether this finding is related to the tax or other characteristics of the states in which households resided.

Table 6. Adoption Model: Estimation Results

Variable	Parameter estimate and [p-value]
Intercept	-1.4919[0.0001]a
Internet taxation	-0.1683[0.0745]
Number of broadband providers	0.0118[0.6101]
Income between $30,000 and $49,900	0.4531[0.0024]a
Income between $50,000 and $99,900	0.7429[0.0001]a
Income $100,000 or more	1.1331[0.0001]a
Race-white	0.2905[0.0405]b
College graduate	0.3525[0.0009]a
Age 34 to 49	-0.2239[0.0759]
Age 50 or older	-0.3316[0.0217]b
Children	0.1318[0.3894]
Household size	0.1241[0.3894]
Occupation-professional	0.2610[0.1409]
Occupation-clerical, sales, or technical	0.2098[0.2867]
Occupation-blue collar	0.2638[0.3879]
Occupation-other	0.0212[0.9086]
Rural location	-0.3234[0.1892]
Suburban location	0.0983[0.3406]
Number of observations	901
1-LogL/Log0	16.2800

Source: GAO.

[a] Significant at the 1 percent level.

[b] Significant at the 5 percent level.

Appendix IV. Additional Communications Technologies

Based on our conversations with stakeholders, and our own research, we identified several emerging technologies that could further the deployment of broadband service.

Broadband over Power Lines

Broadband over power lines (BPL) is an emerging competitive source of broadband to the home. BPL transmits broadband by using existing electric distribution networks, such as the wires that deliver electricity to consumers. Although there are a few commercial deployments, most BPL efforts are currently at the trial stage. Trials and commercial deployments range across the urban-rural landscape, from Cullman County, Alabama, to Cincinnati.[7] Currently, BPL can provide upstream and downstream speeds of 3 million bits per second (Mbps), and next generation equipment is being developed to provide speeds of 100 Mbps.

Industry stakeholders have identified several concerns with BPL service. First, while traveling across the electric network, BPL can emit signals that interfere with other users of the spectrum, such as amateur radio and public safety. The Federal Communications Commission (FCC) has taken steps to document, mitigate, and alleviate this potential problem. Second, some stakeholders also expressed concern that, due to the age or condition of the electric network, providers in some areas would be unable to transmit Internet data at high speeds. Finally, some stakeholders expressed varied opinions about the feasibility of BPL to bring broadband service to rural areas. Some stakeholders were optimistic about BPL's ability to serve these communities, while others expressed skepticism, pointing out that overcoming BPL's distance limitations would require more equipment and additional costs.

Wireless Fidelity (Wi-Fi)

Wi-Fi-enabled wireless devices, such as laptop computers, can send and receive data from any location within signal reach—about 300 feet—of a Wi-Fi-equipped access point. Wi-Fi provides data transmission rates, based on the current transmission standard, of up to a maximum of 54 Mbps,[8] which is shared by multiple users. Wi-Fi equipment and services are based on the 802.11 series standards developed by the Institute of Electrical and Electronics Engineers (IEEE) and operate on an unlicensed basis in the 2.4 and 5 GHz spectrum bands. Several stakeholders we spoke with said that Wi-Fi service complemented, rather than substituted for, other broadband services.

The number of areas that can access Wi-Fi service, known as "hot spots," has grown dramatically and, according to one equipment manufacturer, may exceed 37,000. Wi-Fi hot

[7] BPL companies, such as Current Communications, offer the ability to improve the monitoring, detection, and management of the electrical distribution network through improved communication capabilities inherent in BPL equipment. These features increase the attractiveness of BPL to electrical companies, as the companies receive the benefits of improved network operation and potential revenues from broadband service.

spots include such diverse entities as airports, colleges, retail establishments, and even entire towns. Increasingly, municipalities are planning or deploying larger area or citywide hot spots; some municipalities considering or deploying a Wi-Fi network include Atlanta, Philadelphia, San Francisco, and Tempe, Arizona. While Wi-Fi service is widely deployed in urban and suburban areas, some stakeholders identified a few problems with the service. Because Wi-Fi hot spots operate in unlicensed spectrum, interference can be a problem. Several stakeholders we spoke with mentioned congestion or limited distance capability in Wi-Fi as a potential limitation of the service.

Worldwide Interoperabiity for Microwave Access (WiMAX)

With WiMAX service, the distance covered and data transmission speeds can exceed those found with Wi-Fi service. WiMAX can provide data transmission speeds of 75 Mbps with non-line-of-sight service—that is, the signal can pass through buildings, trees, or other obstructions—or up to 155 Mbps with line-of-sight service. In a non-line-of-sight environment, WiMAX can provide service in an area with a radius of 3 miles or more; in a line-of-sight environment, WiMAX can provide service up to approximately 30 miles. WiMAX equipment and services are based on the IEEE 802.16 series of standards and operate in unlicensed and licensed spectrum.

WiMAX networks are being deployed on a trial commercial basis, but some challenges remain for further deployment. More than 150 pilot and commercial deployments of WiMAX networks are currently in use. Because of its greater capabilities in terms of distance and speed, WiMAX can extend wireless broadband to less densely populated communities, where wired solutions may be more expensive to deploy. Stakeholders we spoke with serving smaller, less densely populated areas indicated that they were testing or interested in WiMAX to serve their communities. However, concerns have been raised about spectrum availability, interference, and the ability of different manufacturers' equipment to support the same level of broadband applications. FCC has several initiatives under way to increase the availability of spectrum for WiMAX services. While the WiMAX Forum Certification Lab certifies WiMAX equipment, the standard allows manufacturers of equipment various options, such as different levels of security protocols, and thus, not all equipment may support the same level of service, such as carrying voice over the Internet (VoIP) and security.

Third Generation (3G) Cellular Broadband

Recently, several major commercial wireless companies have introduced broadband service based on advances in cellular technology and data protocols. Focused primarily on the business customer and more expensive than cable modem and DSL services, 3G services permit consumers to receive broadband service while mobile. 3G services typically provide data transmission speeds of 400 to 700 kilobits per second (Kbps). There are two competing technologies: EVDO service, introduced by Verizon and Sprint; and HSDPA, introduced by Cingular. Currently, Verizon Wireless reports that its service is available nationally in 181

[8] Discussions are underway for newer standards for Wi-Fi that would dramatically increase the transmission speeds.

major metropolitan markets, covering approximately 150 million people. Sprint reports providing EV-DO service in major airports and business districts in 212 markets, covering approximately 140 million people. For HSDPA service, Cingular reports that its service is available to nearly 35 million people in 52 communities. Industry stakeholders expressed concerns about the ubiquity of service, data transmission speeds, and the monthly costs associated with 3G service. Opinions varied as to whether cellular broadband services would be a competitive threat, or a complementary service, for consumers of other broadband services.

Fiber to the Home (FTTH)

FTTH provides a high-speed, wire-based alternative to traditional cable and telephone networks. According to the FTTH Council, as of September 2005, 2.7 million homes were passed by fiber and over 300,000 homes were connected to fiber in 652 communities in 46 states. Stakeholders expressed concerns about the high cost associated with deploying FTTH, and also that FTTH deployment was concentrated in urban and suburban communities, or in newly developed communities (known as "greenfields").

Appendix V. Comments from Industry Participants

We provided a draft of this report to several associations representing industry trade groups and state and local government entities for their review and comment. The following associations came to GAO headquarters to review the draft: Cellular Telecommunications and Internet Association (CTIA), National Association of Regulatory Utility Commissioners (NARUC), National Association of Telecommunications Officers and Advisors (NATOA), National Cable and Telecommunications Association (NCTA), National Telecommunications Cooperative Association (NTCA), Satellite Industry Association (SIA), US Internet Industry Association (USIIA), United States Telecom Association (USTA), and Wireless Internet Service Providers Association (WISPA).

Officials from CTIA, NARUC, and NTCA did not provide comments. Officials from NATOA, NCTA, SIA, and USIIA provided technical comments that were incorporated, as appropriate.

USTA officials noted that our discussion of the effects of local franchising on deployment imply that franchise agreements have helped to ensure broad deployment of broadband, but that, in the view of USTA, franchise buildout requirements can deter entry and thus reduce deployment.

WISPA officials expressed concern about our findings regarding the taxation of Internet access and noted that it is important, in their view, that wireless Internet access provided by small providers not be taxed, and in fact, WISPA officials noted that small providers should be provided a tax incentive to encourage investment and expansion in underserved areas. Additionally, these officials expressed concern about the presentation of data on how households currently access the Internet from their homes. WISPA stated that these data understate the importance that wireless access will have toward the goal of universal broadband coverage both within and outside of users' homes. WISPA stated that the report accurately depicts that wireless Internet service providers (WISP) currently hold a minority market share, and WISPA officials note that without certain government policies to foster growth in the wireless industry, WISPs will be at a competitive disadvantage. WISPA officials also expressed concern that the report understates factors that are hindering the growth of the wireless Internet industry—most notably, the need for additional spectrum under 1 Ghz, such as the TV white spaces. Further WISPA noted that the data showing broadband penetration rates in urban, rural, and suburban areas should not be interpreted as indicating that access to broadband is lower in only rural areas. They suggested that differences in broadband penetration rates across these types of locations are not that great and that pockets of areas with no access exist in many areas. As such, WISPA suggests that policy response regarding spectrum availability, USF funding, and Rural Utilities Service be focused on engaging smaller providers that can bring broadband to areas not currently served by the larger incumbent providers

ENDNOTES

[1] Throughout this report, we refer to high speed Internet access over broadband technologies as broadband Internet access.

[2] See GAO, *Internet Access Tax Moratorium: Revenue Impacts Will Vary by State*, GAO-06-273 (Washington, D.C.: Jan. 23, 2006).

[3] While FCC states that its zip-code information is not meant to be a measure of broadband deployment, some parties have used it in this manner because there are no other official data on deployment of broadband across the country.

[4] The 3-mile limit applies to the path taken by the telephone wire, not necessarily a straight line between the central office and the customer's residence.

[5] With fiber feeders, DSL service can be extended beyond three miles from the central office.

[6] FCC defined "advanced service" as exceeding 200 Kbps both upstream and downstream and "high-speed" service as exceeding 200 Kbps in at least one direction, in order to distinguish these from existing data services based on widely available analog telephony and ISDN technology.

[7] Spectrum is a natural resource used to provide an array of wireless communication services. FCC regulates commercial entities' use of spectrum. With unlicensed spectrum, a number of users without licenses share a portion of the spectrum, adhering to certain technological specifications. In contrast, with licensed spectrum, FCC provides entities with a license to use a specific portion of the spectrum.

[8] In the past, companies with less than 250 broadband connections were not required to submit information to FCC through Form 477. FCC officials told us that many of the companies that are now required to report are very small and in rural areas. These officials stated that many of these companies are not reporting and that therefore the data may not fully represent broadband deployment.

[9] FCC requires providers to report on their broadband lines or wireless channels. While this may not exactly equate to subscribers, the number of lines and subscribers is related, and

we use the word subscribers throughout this report as we refer to the 477 filings of companies.

[10] We used survey data from Knowledge Networks/SRI's *The Home Technology Monitor™: Spring 2005 Ownership and Trend Report.*

[11] A very small number of respondents to the survey accessed the Internet over a satellite connection, but none of the respondents reported any other means of wireless access.

[12] We refer to rural areas as areas outside metropolitan statistical areas (MSA); suburban areas as areas within an MSA but not a central city; and urban areas as a central city of an MSA.

[13] Our findings are not substantially different from those of other organizations. Based on 2003 data, the Census Bureau reported that 62 percent of American households had a computer—see U.S. Census Bureau, *Computer and Internet Use in the United States: 2003* (Washington, D.C., 2005). Additionally, the Department of Commerce reported that 20 percent of households—or 37 percent of online households—had broadband service, with DSL becoming increasingly popular. This study also found that broadband service was less commonly purchased in rural areas—see U.S. Department of Commerce, *A Nation Online: Entering the Broadband Age* (Washington, D.C., September 2004). Similarly, using survey data from 2005, the Pew Internet and American Life Project reported that 53 percent of Internet users subscribed to broadband service, that much of the growth in broadband service in recent years arose from DSL subscriptions, and that broadband service was less prevalent in rural areas when compared with broadband subscribership in suburban and urban areas.

[14] In a recent report, we also noted that the 477 data do not provide a full description of broadband services for certain segments of the population, such as Native Americans residing on tribal lands. See GAO, *Telecommunications: Challenges to Assessing and Improving Telecommunications for Native Americans on Tribal Lands*, GAO-06-189 (Washington, D.C.: Jan. 11, 2006).

[15] See FCC, *High-Speed Services for Internet Access: Status as of December 31, 2004* (Washington, D.C., July 2005).

[16] The problems related to tracking data on subscribership versus deployment/availability in Form 477 is not an issue with mobile wireless operators. Because mobile wireless broadband services are designed to be used while subscribers are mobile, those operators are directed to report the zip codes covered by their mobile wireless broadband networks, rather than the zip codes of the billing addresses of their subscribers.

[17] As noted earlier, some households might not be able to actually receive broadband over satellite if they do not have a clear view of the southern sky.

[18] UNEs are physical and functional elements of the telephone network, such as the telephone line, or loop, which, under the Telecommunications Act of 1996, incumbent telephone companies must make available to competitors for lease or purchase.

[19] In particular, we removed satellite providers, removed any companies we determined only provide service to business customers, removed a cable provider if we found that more than 1 of the largest 10 cable providers served the zip code, removed a cable provider if the respondent said that cable does not pass their residence, and removed telephone-based providers if the residence was further than 2.5 miles from the central office that served the respondent's home.

[20] See Tony H. Grubesic and Alan T. Murray, "Waiting for Broadband: Local Competition and the Spatial Distribution of Advanced Telecommunication Services in the United States," *Growth and Change* (2004), 139-165.

[21] We did not contact officials in Alaska and Hawaii, since the survey data from Knowledge Networks/SRI did not include households from these two states.

[22] This program also can provide loan guarantees, but to date, no loan guarantees have been requested

[23] A recent GAO report, *Telecommunications: Challenges to Assessing and Improving Telecommunications for Native Americans on Tribal Lands*, GAO-06-189 (Washington, D.C.: Jan. 11, 2006) discusses how leadership in a community can help to improve telecommunications services on tribal lands. The report provides several examples of tribes addressing the barriers to deployment of telecommunications networks by partnering with private entities, providing technical training, and taking initiative to access federal grants.

[24] "High-income households" were defined as those having incomes in top 25 percent of all households, while "low-income households" were defined as those having incomes in the bottom 25 percent of all households.

[25] We define "young adults" as people between the ages of 18 and 33.

[26] See Debra J. Aron and David E. Burnstein, "Broadband Adoption in the United States: An Empirical Analysis" (paper presented at the 31st Annual Telecommunications Policy Research Conference, Arlington, Va., 2003).

[27] We recognize that our model does not fully include all the variables that would influence the adoption decision. As such, the parameter estimates will be biased. We are unable to assess the possible extent of this bias.

[28] U.S. Department of Agriculture, Office of Inspector General, *Audit Report: Rural Utilities Service Broadband Grant and Loan Programs*, Audit Report 0960 1-4-Te (Washington, D.C., Sept. 30, 2005).

[29] According to RUS officials, demand for the grant program exceeds available funding under the current program requirements.

[30] RUS officials noted that many different technologies were used in these 19 communities, so that the cost per household varied considerably across the grant recipients.

[31] For rural ILECs, the cost of service is based on "embedded costs"—or the historical costs of infrastructure, which is used to provide a variety of communications services. For nonrural ILECs, the cost of service is based on the forward-looking costs of providing only certain telecommunications services.

[32] See GAO, *Telecommunications: Strong Support for Extending FCC's Auction Authority Exists, but Little Agreement on Other Options to Improve Efficient Use of Spectrum*, GAO-06-236 (Washington, D.C.: Dec. 20, 2005).

In: Maximizing Broadband Services to Rural Communities ISBN 978-1-60692-846-2
Editors: V.B. Haynesworth, A.R. Harrison, pp.151-251 © 2009 Nova Science Publishers, Inc.

Chapter 6

SUBCOMMITTEE HEARING ON MAXIMIZING THE VALUE OF BROADBAND SERVICES TO RURAL COMMUNITIES. SUBCOMMITTEE ON URBAN AND RURAL ENTREPRENEURSHIP COMMITTEE ON SMALL BUSINESS, UNITED STATES HOUSE OF REPRESENTATIVES*

U.S. Government Printing Office

* Excerpted from U.S. Government Printing Office, Washington. One hundred and tenth Congress, First Session, 34-832 PDF, dated May 9, 2007.

Michael Day, *Majority Staff Director*
Adam Minehardt, *Deputy Staff Director*
Tim Slattery, *Chief Counsel*
Kevin Fitzpatrick, *Minority Staff Director*

UBCOMMITTEE ON URBAN AND RURAL ENTREPRENEURSHIP
HEATH SHULER, North Carolina, *Chairman*

Rick Larsen, Washington
Michael Michaud, Maine
Gwen Moore, Wisconsin
Yvette Clarke, New York
Brad Ellsworth, Indiana
Hank Johnson, Georgia

Jeff Fortenberry, Nebraska, *Ranking*
Roscoe Bartlett, Maryland
Marilyn Musgrave, Colorado
Dean Heller, Nevada
David Davis, Tennessee

Wednesday, May 9, 2007

U.S. House of Representatives,
Committee on Small Business,
Subcommittee on Urban and Rural Entrepreneurship
Washington, DC.

The Subcommittee met, pursuant to call, at 10:00 a.m., in Room 2360 Rayburn House Office Building, Hon. Heath Shuler [Chairman of the subcommittee] presiding.

Present: Representatives Shuler, Clarke, Fortenberry, Musgrave, and Davis.

OPENING STATEMENT OF MR. SHULER

The Committee met, pursuant to call, at 10:00 a.m., in Room 2360 Rayburn House Office Building, Hon. Heath Shuler [chairman of the subcommittee] presiding.

Present: Representatives Shuler, Clarke, Fortenberry, Musgrave, and Davis.

Also Present: Representative Fallin.

Chairman Shuler. Good morning, everyone. I am pleased to call this hearing to order. This is the Subcommittee on Rural and Urban Entrepreneurship, the first hearing of the 110th Congress.

I would like to welcome the Members of the Subcommittee, the distinguished witnesses, and our guests.

Rural America is the home of many different kinds of small businesses. In Western North Carolina, there are thriving high-tech businesses, small manufacturers, and family farms. What all these businesses have in common is the need to stay connected with their customers, their suppliers, and the information that they need to run their businesses.

Today's hearing will focus on the challenges for providing reliable, affordable broadband access of the rural small businesses. Experience has shown that broadband can bring economic revitalization to small towns by creating clusters of small businesses. Broadband service can also help farmers and farm-related businesses control costs and optimize production. This technology can provide real-time access to whether reports, fertilization guidance, and livestock tracking.

Farming communities must maximum the use of high-tech Internet access to ensure further development. I am concerned that many of our country's rural and agricultural-based communities are not yet fully realizing the potential of value of broadband services to their economies. As more of these communities gain highspeed access to the Internet, the next challenge is to help them use the access effectively to help create jobs and sustain growth.

There are many debates going on right now about broadband policy. We must ensure that the needs of the rural, small businesses are taken into account whether local, state, or federal governments act to change the broadband marketplace.

During this hearing, I hope that we will begin a dialogue that will help make this happen.

I am very pleased that we have two expert panels here this morning and I look forward to hearing their testimony.

I would like to recognize the Ranking Member, Mr. Fortenberry, for his opening statement.

Opening Statement of Mr. Fortenberry

Mr. Fortenberry. Thank you, Chairman Shuler, for your remarks and for scheduling this important hearing today and thank you all for, especially to our witnesses, for your willingness to appear before us today.

This Subcommittee is the only Committee tasked with the exciting responsibility of encouraging entrepreneurship. Entrepreneurial small businesses are the creators of most new jobs in our country and many Americans are rethinking the old concept of work in favor of being their own boss and bringing new products, innovations, and services to the marketplace.

This is also a very decided trend among younger people and I do believe that the work of this Subcommittee can help remove some of the barriers to creating a more entrepreneurial society.

As you all are aware, we are here to discuss broadband Internet access and its importance to rural America. Broadband provides an important gateway to innovation and the tools for adapting to the ever-changing marketplace.

A study last year by the Massachusetts Institute of Technology documented that communities which had broadband availability enjoyed more rapid growth in employment. Across rural America, businesses in health care, retail, and the agriculture sectors are re-alizing important innovation through the growth of advanced broadband services.

America, however, has been slower than other nations to adopt this technology. Currently, 36 percent of households have broadband access, but the percentage of broadband usage is actually lower in rural parts of the country. According to a General Accountability Office report, the GAO report, the adoption rate of broadband services in rural areas is only 17 percent, much lower than the national average.

In my own State of Nebraska, we're fortunate that parts of all 93 counties have some form of broadband Internet access, however, nearly 400,000 Nebraskans live outside population centers making it more likely that they cannot access this vital service.

In today's hearing, we will hear about the potential benefits of having more competitive services for broadband in rural America and review some of the barriers that stand in the way of such development. In addition, we will review state efforts to pave the way to an expansion of access and I'm particularly interested in how one state, Kentucky, has become a national leader on this issue. Their effort demonstrates that there is currently no strong definition of what is an unserved area and illustrates the importance of creating a methodology for defining what areas of the nation are unserved or under served.

Improving the climate for entrepreneurs will depend, in part, on a more nuanced effort by interested federal agencies to answer these questions and find out what areas of the country need the most attention. Again, thank you, Mr. Chairman, for calling this important hearing and I look forward to the testimony.

Chairman Shuler. I ask unanimous consent that the record be open for five days for Members to submit their statement.

Hearing no objection, so ordered.

Our first panel, this morning, I'd like to introduce, the Commissioner, Mr. Adelstein. He became the Federal Communications Commissioner on December 3, 2002 and was sworn in for a near five-year term on December 6, 2004. Before joining the FCC, Commissioner Adelstein served for 15 years as a staff member of the United States Senate. For the last seven years, he was a senior legislative aide for the United States Senate Majority Leader, Tom Daschle, where he advised Senator Daschle on communications, telecommunications, financial services, transportation, and other key issues. Commissioner Adelstein was born and raised in Rapid City, South Dakota and now lives in Washington, D.C.

Mr. Commissioner, thank you for being here and we look forward to your testimony.

Statement of the Honorable Jonathan Adelstein, Commissioner, Federal Communications Commission

Mr. Adelstein. Thank you, Mr. Chairman, and Congressman Fortenberry, for inviting me to testify this morning. I'm especially honored to be here for your first hearing of the Subcommittee and I'm really pleased that the first subject that you're taking on is deploying affordable high-speed broadband everywhere in this country. I think we need to make broadband the dial tone of the 21st century.

As you mentioned, I grew up in South Dakota and my family business out there was building roads and bridges that helped grow our state by connecting the physical connections between communities in the state, but broadband networks now are bringing people together in ways that my engineer grandfather never could have imagined. As a bridge builder though, I think he would have understood the importance of broadband for commerce.

Broadband is especially critical to economic future of rural America where it can connect businesses to millions of new customers, facilitate telecommuting, and increased productivity, and I could go on and on and I do at length in my testimony. But what's really at the heart of it is that broadband can restore the sense of opportunity that first inspired Americans to settle the frontier and provide hope to our young people who want to stay in the communities where they grew up or return, as we discussed beforehand. They want to come back home. They need to know that there's going to be economic opportunities that await them without a kind of broadband network that would make it just that much harder for them to get back home and stay there.

Especially in an age of global competitiveness which we'll hear about more from the panelists, no matter where people live, we've got to tap their full potential. There are good lessons to draw on. You'll hear many from the next panel, but I'm extremely concerned that as a nation, we're failing to keep pace with our chief global competitors. Citizens of other countries are getting a much better broadband deal. More megabits for less money. It's a productivity problem for economy and we've got to do better.

Some argue that we've fallen in these international broadband rankings precisely because we are such a rural country. Well, if that's true, we should redouble our efforts and address that issue head on because we certainly want to maximize rural economic development and our overall economic growth. We need to prevent outsourcing of jobs overseas by promoting the insourcing of jobs by U.S. companies within our own borders.

I'm concerned that the lack of a coherent broadband plan is one reason that we're falling behind. It's an urgent priority to create a comprehensive national broadband strategy that targets the needs of every part of this country, including rural America. It's got to incorporate benchmarks, deployment time tables, and measurable thresholds to gauge the progress that we're making.

We need to set ambitious goals that aim at true high-speed broadband. One first important step is to update our current anemic definition of broadband that we have at the FCC of just 200 kilobits in one direction, something that's more akin to the kind of speeds they're getting overseas that will really support video and data services.

We should start by gathering more reliable, specific data than the FCC currently compiles so we can better ascertain problems and develop solutions. I think you're right, Connect Kentucky showed us a great model and there's no reason that we can't do it on a national level what Kentucky was able to do on a state level. We've got to increase incentives as well because the market will be the primary driver of companies that benefit from a stable, regulatory environment.

We must also work to promote meaningful competition which is the most effective driver of innovation. It keeps prices low.

Federal universal service continues to play a vital role in maintaining and improving these rural networks. As voice becomes just one broadband application along with video and voice and data, we need to ensure that universal service evolves to provide a ubiquitous advanced services, a priority the Congress made clear in the Communications Act that was updated in 1996. One major growth engine for broadband, particularly in rural areas is the potential of spectrum-based services, wireless services. We've got to get spectrum into the hands of operators ready to serve at the most local levels. Previous auctions, I pressed for the use of smaller license blocks. I want a balanced facilitating spectrum access for those providers who want to offer service to smaller areas, with giving those larger carriers strategic opportunities to expand their footprints as they need to.

I think we really have an historic opportunity in this upcoming 700 megahertz auction. This is the television spectrum that we're going to re-auction coming up early next year or late this year. That could really facilitate an emergent third, broadband platform, a real national wireless broadband network. To make that happen, I think our auction rules should provide a diverse group of licenses, giving all bidders the chance to win licenses that best match their business plans.

Unlicensed wireless is also part of the rural solution. Unlicensed spectrum is free and in most rural areas, it's lightly used. It can be accessed immediately using widely available technology. We're working to make more unlicensed spectrum available at higher power levels and we're evaluating unlicensed operations in unused TV spectrum bands, the so-called white spaces.

There's a lot more that Congress can do as well. Just a few ideas: providing adequate funding for and properly targeting rural utility service broadband loans and grants; providing tax incentives to companies that invest in broadband in under-served areas; revising better depreciation rules for capital investments and targeted telecommunication services; investing in basic science and research and development for further innovation; and improving math and science education so human resources can continue to fuel technological growth.

Just as roads and bridges paved the way for economic success of rural America in the last century, broadband networks will be a big part of maintaining and restoring the vitality of our rural communities in the future.

Thank you for your leadership on rural broadband by holding your very first hearing on this subject and I appreciate the opportunity to testify and I'm happy to answer any questions you might have.

[The prepared statement of Mr. Adelstein may be found in the Appendix on page 40.]

Chairman Shuler. Thank you, Commissioner. You mentioned some incentives. One was tax incentives, others was depreciation schedule. How can some of the smaller businesses benefit from that with having both a public and private relationship of extending broadband services in rural areas?

Mr. Adelstein. I think public/private partnerships work really well. We'll hear from Connect Kentucky and I believe that one thing the government can do to help out is if companies do really target these areas that are higher cost, it's obviously more expensive to provide broadband to rural areas, but the benefits to the economy are such that it justifies perhaps in having tax credits that can encourage the development of broadband where it might not otherwise be economically feasible.

Similarly, in our U.S. grants and loans have been successful in the past, I'm not sure they've been as well targeted. The House Agriculture Committee held a hearing where we saw that some of the subsidies were going to the wrong places and weren't going enough to the right places. Their own IG found the same of targeting, but there's a place where public and private sectors can work together as well to try to give incentives so that the market works better.

Where there's a failure of the market to operate on its own, use a market-based mechanism like a tax credit or a low-cost loan, just to give a little edge that these providers need to make that investment in rural America. I think that will pay back in spades to the overall economy.

Chairman Shuler. How do we help the small businesses? Once they have, there's broadband access to particular rural areas, how do we encourage from a community standpoint, how do we encourage a small business, what kind of learning curve, what type of progress should we be making to really encourage the small businesses, up-starting businesses to actually come back to the rural communities that had been going to the larger cities?

Mr. Adelstein. You know, it's something that we haven't done a lot of work on in the FCC, frankly. We probably should be doing more thinking about that, but I keep referring to

Connect Kentucky because we've been reading their testimony, thinking about what they did. What an outstanding job of educating small businesses and others about what is possible and the idea has been out there for a while about demand aggregation. You get these public/private partnerships that go out and educate small businesses. There's a lot of small businesses that don't know now how much they can benefit from this kind of activity.

I think that Lee Terry talked about, Congressman Terry from Nebraska, talked about how one business went from three employees, it was a meat business, to 50 when they learned that they could sell their products over the Internet and they had a broadband connection to do it with. But they hadn't thought of that until somebody came to them and explained it. So these organizations, the local Chamber of Commerce, connecting with local government authorities and state government authorities, there's no reason the federal government can't help as well, really can make an effort to educate these businesses about the possibility, so that they can thrive in these smaller communities.

There's all kinds of untapped potential and of course, the workers in rural America are second to none anywhere in the world and in this country. They're stable. They're reliable. Companies are finding that, but they can't use them if they don't have a broadband connection because a lot of the skills, call centers, or back office operations require broadband connection.

So obviously, state economic development authorities can try to attract businesses to come out and locate there, but it's important for businesses that exist already to have them educated by these kind of public/private partnerships about how broadband can help their business and then going to the providers and saying look at all these small businesses that want broadband. You've got a customer base here. It's worth it for you to invest.

Chairman Shuler. What one thing could we do as a Congress or two things could we do as a Congress to truly have a much bigger impact in the rural areas?

Mr. Adelstein. I've talked about having a national broadband strategy that targets rural America and there's a lot you can do in Congress. We have a big role at the FCC. I talk about universal service. I talk about wireless. That's all been made possible by legislation that Congress has already put on the books, putting more wireless spectrum out there. Let's make sure we get it in the hands of small providers that will serve local levels.

Universal service, let's make sure it evolves to cover broadband at the appropriate time. Those are in place, but Congress can help us with universal service by providing a stable contribution base. We need a broader base because right now it's pegged to a declining revenue base which is long distance revenues which, as we all know, are declining. And we need to have a broader base and Congress could help with that.

Congress can also help, I think, by making sure that RUS is fully funded and that the program, as it's reauthorized in the Farm Bill, is properly targeted. I worked on that bill in the earlier Farm Bill when I worked for Senator Daschle who was the senior Member of the Ag. Committee and the Majority Leader at the time. We really thought we did a pretty good job. I mean we put it together and we said target under-served areas. Do grants to unserved areas. And I'm really saddened to learn that sometimes those priorities weren't fulfilled in the implementation of the program. So maybe Congress needs to go back. I thought we did a pretty good job writing it, but if you need to beat them on the head and say focus on real rural America, then you need to do so.

There are so many other areas that you can do, I think R&D, the whole innovation agenda that Speaker Pelosi has put forward, contains a lot of ideas that have been around for a long time, and ones that she's starting to implement. The idea of math and science education is critical so we have the basics that people can become technologically proficient in. R&D, basic R&D funding. We've seen R&D funding has gone up, but it's been largely military or health-related and you don't see basic R&D for science. It's actually fallen behind. So we need to redouble our efforts on basic R&D funding.

There's a lot that can be done. The tax credits I talked about, depreciation rules. Those are some of the areas that Congress can help us, but I think the FCC has a big role with what Congress has already given us, making sure that we do more to come up with this national broadband strategy, in conjunction with Congress so that we don't leave rural America behind.

Chairman Shuler. The chair will now recognize Ranking Member Mr. Fortenberry.

Mr. Fortenberry. Thank you, Mr. Chairman, and thank you, Chairman Adelstein, for coming today. We appreciate the opportunity to visit with you and appreciate your insights. I also happened to be on the Subcommittee, on the Ag. Committee, that oversees the Rural Utility Service and we had a very exciting and interesting hearing on this very topic several weeks ago.

It's fairly clear that there are some, to put it mildly, dilemmas there. I appreciate the point you made about us not having a coherent broadband strategy nationally, the Rural Utility Service's attempt at providing a component of this brings some dilemmas into play as you mentioned. Are we unfairly subsidizing markets which already had or player competition into marks which have already had substantial private sector investment without governmental subsidy and prioritizing that over under served or nonexistent service in other rural communities.

One of the key findings that came out of that was there is not a clear understanding, as I mentioned in my earlier testimony, who's under served, who is not served, and we at a federal level our first response is we need to find out that question. But as you mentioned in your testimony, the Kentucky model, might point to a different solution to that because it is a smaller scale implementation that apparently begins with some simple concepts in terms of just informing the marketplace, particularly small businesses about the potential opportunities they have in using broadband to expand their services, thereby creating a more natural momentum in the private sector.

But nonetheless, I want you to comment on that. A lack of coherent or unpack your statement a little bit more, a lack of coherent broadband strategy or lack of coherent broadband plan would mapping the situation in the country be an aid in that regard or is the technology too variable, too fluid, and by the time we would get this done potentially at the federal level, having already shifted and changed. Would that be a problem or would that be a potential solution?

Mr. Adelstein. I think mapping is an essential solution. I think that it's—we've got to do a far better job of broadband data gathering. The first step of any national broadband plan is to map out what we've got now so we know where the problems are and we can better develop solutions. GAO has been very critical of our efforts so far to assess broadband data. We need more granular data. We don't have data on the local level. We need a better definition of broadband. We're looking at 200 kilobits in one direction. That might have been good back when we invented it in the '90s, but it's broadband any more because it doesn't carry critical services like video or telemedicine or on-line learning programs.

What we're doing now are FCC data, looks at the zip codes. It says if you have one person in a zip code that has broadband, you've got broadband throughout the state, throughout that zip code. But that's just not the case. One legislative idea that's been floated I've heard, is to go to the nine-digit zip code, rather than the five-digit zip code and really getting a more localized sense of where there is and isn't broadband. We've got to have better data.

We recently as a Commission all unanimously voted out an improvement. Commissioner Copps and I have been talking for years about how we get better data on availability, more localized data. We voted out an order asking questions about how we're going to improve our data collection by looking at more demographics of subscribers. Even if we need to do surveys to find out rich, poor, race, any other, male, female, we need to know more about the demographics, where they are, what they're doing with the service.

I think this effort that we've launched on the FCC is overdue, but it certainly is welcome and I'm glad that we're doing that. It's going to take a while before that data comes in. And any guidance Congress can give us, for example, if they tell us to go out there and gather data on a nine-digit zip code and give us some funds to go out there and map that, I think it would be a good investment because in the end, that will be the first basis of the real national broadband strategy, where we go, how we deal with the problems in rural America.

Mr. Fortenberry. Do you interface with the Rural Utility Service on that very question or are you in separate buildings and not in communication on the issue?

Mr. Adelstein. We're pretty close. We talked about coordinating. We did some—I think we could do more. Again, nobody has got the data. They look at it when they get an application, they get a loan application or a grant application in and they look at the situation there, but we don't provide that much data because we don't have a lot on a local level. It's not like Kentucky.

The example of Kentucky, our survey found that 96 percent of the people in Kentucky have broadband. They went and looked at the more granular level, it turned out it was only 74 percent. So we were way off. The situation wasn't nearly as rosy as it appeared from our zip code data. So there's not a lot we can offer RUS in terms of how to target it. I wish we could, but they're it more on a case-by-case basis and I think as they look at the case, they get more granular data about that market than we could ever —

Mr. Fortenberry. One of the findings was that the loan application itself was the determining factor as to whether or not the area had broadband. So I think that some maturity of our process by which we come to a better understanding of where services truly are needed in order to more precisely target our limited funds as a federal government is important and prudent. So I appreciate that offer that you just—I'll take it as an offer and we'll think through that.

One other quick point, Mr. Chairman, Mr. Adelstein mentioned the—we're talking primarily about the impact on business opportunity in rural America, but the impact on the delivery of health care and education, that this new highway, digital highway can provide, is extraordinary and I think we've got again a magnificent opportunity here to continue to measure the impacts that this technology can have in rethinking the delivery, not only of business services, but all types of social services as well that are essential well being of our country.

So I appreciate that observation. It's a very important one, as well. Undersecretary Dorr who is in charge of USDA's Rural Development pointed to this factor as being the most significant in marketplace change that we could impact on behalf of good rural development, advancing the access of broadband throughout the nation. So thank you for your input.

Chairman Shuler. I'd like to commend the Ranking Member for his comments. With education and health care that is a vital part of what the access to broadband can help. In our District alone, we have 16 hospitals and now all of which are connecting to the major regional hospital in our area and it's an overwhelming amount of savings and costs that they can actually cut in the health care industry. It's going to be tremendous for just once again not having to duplicate services alone, to be able to access the information of the patients' medical records in order to better access the information more readily available through the use of the broadband technology. So I comment you for that.

The chair now recognizes the gentlewoman from New York, Ms. Clarke.

Ms. Clarke. Thank you very much, Chairman Shuler, for holding this very important hearing to explore how broadband services can revitalize rural economies. Some may be wondering why I'm here since my District is not a rural community. I'm from Brooklyn, New York. However, the Chairman's District and my District share one common factor, that is inadequate broadband service. I fear that my District with its dichotomy of socioeconomic diversity, which has an extremely affluent part of the District where broadband is really not a challenge, but then has a very under- served end of the District where the socioeconomics have not lent to real access to broadband for so many years won't survive this new information age if we do not make available high Internet access.

There are countless stories from many people who live in low-income neighborhoods about how they do not have or are unable to receive access to high-speed broadband services. I personally would like to see more aggressive efforts to increase Internet access in disadvantaged communities.

And so my question to you today, Mr. Adelstein, is although broadband has become increasingly available for people of modest incomes, it has not reached those living at the lower end of the income scale. According to the most recent report by the Pew Internet and American Life Project, only 21 percent of households with an income of $30,000 or less had a broadband connection at home in the Year 2006, while 68 percent of households that earn over $75,000 a year, had a home broadband connection.

I would like to know have you done an assessment of what can be done to get more low-income households to obtain broadband connection?

Mr. Adelstein. Well, the sad answer is we really haven't done as much work on this as we should. There's a real problem with pricing in some of these areas as well, even if people do have access, the prices haven't come down and we're talking about what's available overseas. I was talking to some of my French counterparts and throughout France for $40, you can get 100 channels. You can get broadband up to 20 megabits and your phone service, for $40 a month. What's happening to our economy here, the money is being demanded, $40 just to get a broadband connection. That puts it out of reach of a lot of people, even if it is available to them and you end up with statistics like that which the Pew data shows.

We need to do a better job, I think of promoting competition. It's a little easier, frankly, to get competition into a real dense area like that than it is to get it out in some of the rural areas and competition should be the driver of lower prices and better quality service, but when you see that they go around some of the low income areas and the deployment isn't nearly as good, it's a real problem.

Another issue, of course, is lack of computers there. In some countries, they're actually giving computers out to their citizens, low-cost computers, and then allowing them to hook up that way which makes a lot of sense, because without a computer, you're left behind in this age. So there could be a digital divide not only in rural America, but between economic strata, which is another area I think we talked earlier with Ranking Member Fortenberry, about the importance of broadband data. That's an area that we should also get the cuts and figure out on demographics, high income, low income, where that penetration is taking place. I think it's especially important that we level that out because there's nothing ultimately that will level out our economic life than having access to technology. And if people are left behind on that, then they're going to continue to not be able to participate fully in our economic life as a country and our overall economic growth will suffer.

Ms. Clarke. And so you believe that government has a role in basically promoting competition with regard to that? For instance, in New York City, while you may not have as many households that have computers in them, we have made sure that many of the public libraries and public facilities within the communities have that available. They just don't have access to broadband because as you said, the competition makes it cost prohibitive. Do you see a role that we can play in encouraging that competition and what do you think can be done in the short term to help these communities to really access the information highway?

Mr. Adelstein. We have to do more about competition. That was the real focus of the Communications Act of 1996. It's critical because it's the essential input into the economy. I mean it's a time of great change, of course, in these services. You see new services emerging. People talk a lot about convergence and new players are coming in. The loss of competition after the creation of the Act, is being supplemented to some extent by other forms of competition, although the pace is unclear. You do have cable fighting it out with the telephone companies which is a battle between facilities based providers. It should be helping us, but you know, we're seeing consumers embracing these new technologies. I think the challenge is how to function in this new market.

We need better data collection to start with, better analysis of the facts, and we need to promote healthy competition by leveling the playing field, but not blinding ourselves to where competition isn't sufficient to safeguard consumers.

Ms. Clarke. Thank you very much, Mr. Chairman. I yield back. ChairmanSHULER. The chair will not recognize the gentlewoman from Colorado, Ms. Musgrave.

Ms. Musgrave. Thank you, Mr. Chairman, along with Ranking Member Fortenberry. I was on the Ag. Committee when we were talking about this issue and when the mapping issue came up I asked the question about mapping and I was told that it would immediately be obsolete and not have very much use at all. And quite frankly, I'm encouraged by your comments. We need better analysis. We didn't even talk about a nine-digit zip code in that hearing.

But what about mapping? Could you elaborate on that and could you also address the issue of obsolescence that was raised in that hearing, please?

Mr. Adelstein. Well, the marketplace is evolving rapidly, so obviously there's going to be changes, but right now we don't have a good picture at all. I mean to say you're trying to navigate a world with no map at all versus having an old map. I'd rather have an old map than no map. And that's kind of where we are now.

I think that the situation isn't changing that fast. One of the concerns about providers is their proprietary data. They don't want that out there. There's been a lot of hesitation about it and I think you can hear from the Connect Kentucky folks, but I understand it took some cajoling to get providers to provide this data and it may require Congress also to provide us exemptions from the Freedom of Information Act so that their proprietary data can be protected and they feel comfortable giving it to us. Because ultimately, it's in everybody's interest that this map take place.

I know that, for example, Verizon has talked about the importance of getting better data. I mean these companies are beginning to understand that they need to know where the gaps in the marketplace are as well and to know what's available and what's not and it would help RUS and it would also help the FCC. I mean I think all policy makers would benefit. And that's sort of the building block, the basic foundation of a broadband strategy is knowing where we have a problem and targeting solutions appropriately to them.

Ms. Musgrave. How difficult was that cajoling and how much assurance do they need?

Mr. Adelstein. The Connect Kentucky folks might know more about it than me. I certainly think that I've encountered a lot of resistance as I've talked to providers about whether they're willing to share more data with us and we've put out our broadband data notice asking companies for their input on what kind of data we should ask for. I've been very aggressive in saying we want all this kind of data, what do you think of that? We haven't put out the actual report yet, but we've asked them about what kind of report we should put together. And we'll see what kind of response they get. Hopefully, they're beginning to learn that we have a problem in this country and that we need that data in order to solve it.

Ms. Musgrave. Do you think the Universal Service Fund is doing what it needs to do to get affordable broadband to rural America?

Mr. Adelstein. It certainly is the basis, I think, of getting broadband out to rural America. Even though Universal Service doesn't directly support broadband, we have a no barriers policy that allows it to subsidize networks that can carry broadband traffic. So I think we wouldn't see the kind of broadband we are seeing in rural areas without Universal Service and going forward, we need to keep that on a solid basis if we're going to continue to have rural America connected.

I think one basic element of a national broadband policy is a strong, stable Universal Service Fund that properly targets under served areas and ensures that high-cost areas have broadband every bit as available to them as other parts of the country.

Ms. Musgrave. Well, this is very much on my mind. I just worked very hard on getting health care for Veterans in rural Colorado and when we met with communities that were very eager to have this very progressive communities, the issue that the VA brought up was broadband. And so that had to in place before we could get this satellite clinic in rural Colorado and again, Ranking Member Fortenberry has talked about education and health care. And I have communities in the rural part of my District that there's no way you're going to get a specialist out there. We have nurse anesthetists that have to fly to various hospitals in Kansas and Colorado, but for diagnosis and other things the telemedicine works wonderfully.

It's what rural communities have to have for health care, for the education needs. We have many rural schools in my District and some of them have less than one hundred students K through 12. So they face challenges, but yet those students deserve a quality education and they utilize every means possible to make sure that they get it.

Well thank you, for your testimony.

Mr. Adelstein. I just might add that part of Universal Service of course, is the E-Rate which funds schools, libraries and health care, rural health care facilities for this. We recently, it will be of interest to a lot of you, that we recently came up with a rural health care program that is going to offer $60 million in test projects for rural health care and we're getting applications from a lot of your states asking how they can improve the delivery of rural health care through telemedicine. It's really an incredible lifesaving application.

From my home State of South Dakota, they came in and hit me up on it and they were talking about how there was in Parkston, South Dakota, there was a woman who gave birth at 26 weeks. It's a very premature baby and it was in the middle of a blizzard. You don't have to worry about that, Congressman Shuler, as much as some of us do, Congressman Fortenberry and Musgrave. We've got to worry about those blizzards. They're trapped in that. They couldn't get out by helicopter. They couldn't get out by road. They were in this little clinic and there wasn't a doctor there that knew it, but they had a specialist at a Sioux Falls hospital that was in connection, because they had a broadband connection in Parkston and thank God they did, he was able to guide them to the ventilator and basically save this baby's life until it could be transferred to a bigger hospital.

These kind of lifesaving applications are critical and I think that the FCC needs to continue to make efforts like this to test out how we can expand the rural health care program and that's an area that Congress could look at expanding as well.

Ms. Musgrave. Thank you very much. And Mr. Chairman, I do have a written statement that I would like to submit for the record. ChairmanSHULER. So ordered.

Ms. Musgrave. Thank you.

Chairman Shuler. The chair now recognizes the gentleman from Tennessee, Mr. Davis.

Mr. Davis. No questions.

Chairman Shuler. The chair will now recognize the gentlewoman from Oklahoma, Ms. Fallin.

Ms.FALLIN. Thank you, Mr. Chair. I don't know that I have a particular question, but as I was listening to the discussion on rural health care and education and telemedicine and we even use the Internet and broadband for weather delivery service in Oklahoma through a weathernet service that we have, but I had a question about the Universal Service fee and the money that the states have access to expand broadband. Can you explain how the fee is used and how the states can get quicker access to that?

Mr. Adelstein. The money is collected through long-distance revenues which is sort of a declining base and we're trying to figure out ways of broadening the base. If Congress could help us broaden the base, that would be wonderful. In the meantime, we're trying to think within the context of the Act of how we can have a broader base in the declining base of numbers. So we take this big collection, it's a $6 billion program, including the E-Rate, and we—it's granted to companies that are called eligible telecommunications providers. They apply for this program and a state PUC in many cases will designate whether or not there can be an eligible carrier. And if the carrier is eligible, then we come up with a program to define how much they're paid for subscriber.

One of the concerns recently has been that competitive providers are coming in and the money is kind of ballooning because a lot of them happen to be wireless companies and they come in and win these funds based on the amount that's being paid to the incumbent. In many cases, there's a totally different cost basis. The incumbent, it's a lot more expensive for the LEC to provide in its embedded costs than a new wireless company that comes in and provides services. We're now debating how do we equalize that? Do we give the same amount of support to a new company or do we do it on the basis of their actual costs rather than the cost of the incumbent?

All of these funds can underwrite broadband networks. If you didn't have Universal Service, the high cost of serving these rural communities would basically have those systems deteriorating and they would be antiquated. They wouldn't be able to support broadband network. Right now, broadband itself isn't what we call a supported service which is a particular service that we pay for directly. But we have this policy that allows us to fund networks that can carry broadband and so the Universal Service has been one of the most critical elements allowing these rural local exchange carriers to upgrade their networks so that they can carry broadband traffic and we see that sometimes in some rural parts of the country because of Universal Service, they actually have excellent access to broadband that is even better in some more urban areas that don't have access to Universal Service. It's an interesting kind of a situation. So we've got to make sure that remains. Other areas of rural America are falling behind though. All the studies show that despite some of these really wonderful examples we see, there is still a real digital divide that is getting worse.

And so if everybody is saying the problem with American broadband, the reason we're falling behind is because we're a rural country, we better make sure that we keep Universal Service in place and make sure that it continues to underwrite these broadband networks so that we can compete in international economy.

Ms. Fallin. The reason I was asking that question, at different times, I've just heard some talk back in my State that sometimes we're slow to expend those funds. We need to be spending it but yet we hesitate and delay and I was just trying to figure out why we would even delay spending the money that's available to do that.

Mr. Adelstein. There's sort of a dilemma for us on the federal level because a lot of states are really quick. They say well, it's federal money, it's easy, let's just take it. And they don't think a lot about how they award funds to eligible telecommunications carriers. And maybe your state, I'm not sure exactly Oklahoma PUC, whether they're—what they're doing. It may be out of prudence. Sometimes it's wise to be careful about how those funds are expended and we've tried to give states guidance on making sure they're very careful about how they award ETC grants so that they don't just do it willy-nilly thinking it's federal money, let's just let it go, we have nothing to lose. But they really think about the impact also on the LEC and they think if it's in the public interest for these funds to flow and how companies are going to get ETC status are going to expand their networks.

We get nervous if companies are just taking Universal Service and using it for service they're already providing and getting more for what they're already doing. We want to make sure that whatever Universal Service money they get is used to expand and make sure they truly cover everybody in the service area.

Ms. Fallin. Thank you, Commissioner. Thank you, Mr. Chairman.

Chairman Shuler. Does any other Member have any questions for the Commissioner?

Commissioner, thank you so much for your testimony. An outstanding job. As we all know, and as we look around the Committee, we certainly know that the backbone behind our country is our small business, over 95 percent of our business are small businesses and the more access that we will have to broadband and any way that we can help and help guide the broadband system, we would certainly appreciate your help and your guidance and your direction as well.

So thank you for your testimony.

Mr. Adelstein. Thank you for your leadership, Mr. Chairman. ChairmanSHULER. At this time, we'll have the second panel come forward, please.

[Pause.]

Chairman Shuler. I'd like to welcome the second panel. Thank you for your attendance today and your testimony.

We'll go ahead and get started. I just want to remind the panel that I looked at some of your statements. We'll try to keep it to five minutes, best that we can. I know I'm a rookie at this chairmanship, but I will hold the gavel tight. So we'll try our best to stay within five minutes.

The Ranking Member says that he gets hungry around 11:30, so we're going to try our best to get as much information as we can.

Our first panelist, I would like to introduce Mr. Stephens, obviously from my District of Western North Carolina. Mr. Stephens serves as chairman of the board of directors of the Balsam West FiberNET based in Silver, North Carolina.

For the past year, he has been a board member since the company has been founded in 2003. Mr. Stephens is also an enrolled member of the Eastern Band of Cherokee Indians. Mr. Stephens works for the Eastern Band of Cherokee Tribal Government as planner, economic and community development.

Mr. Stephens, thank you so much for your attendance today. We look forward to your testimony.

Statement of Brandon Stephens, Chairman of the Board, Balsam West Fibernet LLC, Sylva, North Carolina

Mr. Stephens. Thank you, Mr. Shuler, and Chairman Shuler. Chairman Shuler and Congressman Fortenberry and Members of the Subcommittee, thank you for the opportunity to share the story of Balsam West FiberNET.

I'm Brandon Stephens, Chairman of Balsam West FiberNET and an enrolled member of the Eastern Band of Cherokee Indians.

I'm here to represent Balsam West FiberNET and partner members of the Eastern Band of Cherokee Indians of Cherokee, Drake Software of Franklin, and Community Alliance Partners Southwestern Community College, headquartered in Sylva. Our goal and mission is to build economic development and stability by offering affordable high-quality fiber optic infrastructure.

Western North Carolina, Eastern Tennessee and North Georgia suffer from a lack of technology and infrastructure. Affording and obtaining access to quality and reliability in telecom have been great barriers to us. Decades ago, during our area's development, telecommunication carries built a network that satisfied demand for the day and not for the future.

The result was a network that did not have redundancy or backup and was susceptible to outages and was of poor quality. Had other utilities used this same approach, power companies would have built power lines as we purchase electric appliances. Home-grown businesses like Drake Enterprises or Drake Software, co-founder and co-owner of Balsam

West FiberNET rely on broadband connectivity. It transmits several billions of dollars in electronic funds transactions and data with the IRS and its clients.

The tribe and local businesses in the region depend on visitors and their ability to access funds for commerce. These outages occurred and no business was conducted for the tribe unless it was in cash. For Drake, billions of dollars in transactions were potentially jeopardized.

As a result, founding members of Balsam West FiberNET developed a 300-mile in-ground fiber optic network because wireless in our region is not viable. That's because of the terrain. We also had built this network to be at the highest standard, so we could offer services to the largest national carriers after developing this network to take care of our own concerns and secure business in our respective areas.

The founding members of Balsam West FiberNET started focusing on the region. We're an open-access system, meaning we allow unrestricted access of services across our network.

As you consider policy changes, I would urge you to know that right now the government is not promoting competition. Forbearance in copper retirement are the exact opposite.

Open access to all platforms would promote innovation and competition. The policy of closed networks is leading us back to monopolies and furthermore, an open network is enabling us to provide our customers to purchase content from the provider of choice.

Options create opportunity to lower prices and increase quality. The spirit of mountain people is to survive and overcome challenges. That spirit has brought us this unique collaboration between Drake software and the Eastern Band of Cherokee Indians. It is this same spirit that has always been required to overcome barriers that hamper development. BalsamWEST is working to develop regional clustering. A prime example of this is the model of Drake software.

Drake is currently expanding into new communities, developing call centers connected in real time over fiber optic cable. We also recently helped the region's schools, creating a distributed learning network called WNC EdNet. This network connects all the schools together with virtually unlimited capacity on fiber. The schools own their own networks and their own private fiber optics.

BalsamWEST and other local infrastructure owners work together to reduce this expense. We save the schools $60 million and gave them the opportunity to choose their content provider.

We also worked on behalf of the rural hospital systems lowering their costs 96 percent. We also decreased the transmission time of imaging from 30 minutes to 12 seconds. There are some barriers to our future development. We have found that federal financial resources are difficult to obtain. Policies in most programs do not lend eligibility to our communities as we fall through the cracks.

Mr. Chairman, another barrier that we hope to overcome soon with your assistance is supporting connectivity outside of our existing network, first to neighboring counties, such as Haywood and Buncombe Counties in Western North Carolina, then to connect to resources in metropolitan areas in Tennessee, Georgia, and South Carolina.

Thank you for the opportunity to share our story of success and challenges and we hope that we can depend on your support in the future of BalsamWEST Fibernet and our motto is "access to advance."[The prepared statement of Mr. Stephens may be found in the Appendix on page 47.]

ChairmanSHULER. Thank you, Mr. Stephens. The chair would now like to introduce Mr. Shields. Mr. Shields is the founder the board chair of Ygomi, LLC, and I hope that's pronounced correctly, which develops and operates technology-based companies. Mr. Shields has over 35 years of experience. Among the technologies his companies have helped to pioneer is technology that is the foundation for the fulfilling billions of on-line direction requests through such services as MapQuest, Yahoo Maps, and Google Maps.

Thank you, Mr. Shields.

STATEMENT OF T. RUSSELL SHIELDS, CHAIR, YGOMI LLC, OAK BROOK, ILLINOIS

Mr. Shields. Thank you, Chairman Shuler, Ranking Member Fortenberry, and Members of the Subcommittee. My name is Russ Shields. I am chair of Ygomi LLC. It's certainly a privilege to be here today to speak to you.

Ygomi is an Illinois-based company with a 37-year record of building companies that deliver innovative software and services to businesses. We're known for applying technology to improve people's lives. Our subsidiaries SEI, Verety, Connexis, and ArrayComm serve leading corporations.

We have more than 1200 employees across the U.S., Europe, and Asia. We provide solutions in areas such as wireless digital signal processing software, vehicle telematics, and technical support for multi-location enterprises using distributed, U.S.-based call centers.

Today's hearing is particularly relevant to the Ygomi Companies. Broadband technology touches them all. For instance, high speed Internet availability in rural areas has allowed us to create a new business solution for one of our customers. Our subsidiary, Verety, now takes drive-through orders remotely in North Dakota for a number of McDonald's restaurants, helping to improve speed of service, order accuracy, and customer satisfaction. Verety can deliver this high-quality service because broadband availability gives us access to people working from home. We have a workforce that includes farmers, stay-at-home mothers, retirees, people with disabilities, and people who care for elderly or disabled family members.

Our employees like the no-commute savings, and the convenience of flexible work shifts. We expect the number of work-at-home employees to increase dramatically in the years to come. We provide each work-at-home employee with a computer, a DSL connection, paid training, and web and phone base support. Our employees and their families can use the computer and Internet connection for themselves when they're not working.

Two-thirds of Verety's work-at-home employees did not have broadband for their families before they came to work for us.

Thomas Friedman's book, The World is Flat, mentions our effort for McDonald's in North Dakota. I believe that Friedman's vision of an efficient, interconnecting flat world is becoming truer every day. Broadband services and voice-over IP gives smaller, more isolated communities access to the world. Companies in the Telework Coalition, like Verety, are helping to realize the benefits of broadband deployment in some rural areas. The same can be done in other areas of the U.S. if the proper incentives are provided. But it takes more than just broadband. It requires a new way of thinking about the workplace and innovation.

We encourage the deployment of broadband in rural areas to ensure that the Internet is available to everyone, no matter where they live. We work with organizations like the Telecommunications Industry Association to promote access to affordable broadband, to minimize regulation, and to maximize the power of a competitive market. We believe in the value of broadband to improve government services, public safety, education and health care.

Increased global competition requires a more flexible labor environment. As a privately-held company, Ygomi can take a long-term approach to profitability. It lets us think flexibly about technologies and applications that will be needed in the future. But we still face challenges. To succeed, we must be responsive to new and evolving employee needs and attitudes. Companies that work on emerging technologies can build businesses in rural areas need favorable environments and incentives. Policies should encourage investment in new and diverse communications technologies in rural areas.

I comment you and your staff for holding this hearing and for your efforts to extend broadband in rural America. Thank you for the opportunity to testify today.

[The prepared statement of Mr. Shields may be found in the Appendix on page 76.]

Chairman Shuler. Thank you, Mr. Shields. The chair now recognizes Mr. Russ Kremer. Mr. Kremer is the president of the Missouri Farmers Union. He is a cooperative business developer who owns and manages a diverse five-family farm.Mr. Kremer is also the president of Missouri's Farmers Union Services, a partner of U.S.A. Broadband LLC, an organization with a mission to provide affordable broadband communication services to all rural residents of the United States.

Mr. Kremer, thank you for your testimony.

STATEMENT OF RUSS KREMER, PRESIDENT, NATIONAL FARMERS UNION, JEFFERSON CITY, MISSOURI

Mr. Kremer. Thank you, Chairman Shuler, and Congressman Fortenberry, and Members of the Subcommittee. Thank you for this opportunity to testify today. My name is Russ Kremer and I am a family farmer and president of the Missouri Farmers Union.

Today, I am here on behalf of the National Farmers Union, our nationwide organization, representing family farmers, ranchers, fishermen, and rural residents. I appreciate the opportunity to highlight the importance of accessible and reliable broadband service to the farmers and producers in communities in rural America.

The future of rural America does depend on high-speed access to the Internet. In 2005, the National Agricultural Statistical Service conducted a study on farm computer usage and ownership. While the results showed that 51 percent of U.S. farms had Internet access, further investigation uncovered that dial-up was the most common method of accessing the Internet with 69 percent of U.S. farms. It is encouraging that more farmers and ranchers gained computer accessibility each year, either through ownership or leasing of computers or other community programs, however, it is alarming that the vast majority of them must do so at the slowest connection speed possible in accessing the most uncommon means of telecommunications.

NFU supports efforts to provide competitively priced high-speed broadband Internet access for rural America. We urge collaborative efforts and public/private initiatives that leverage Internet-based technologies and use the Internet to improve communications, reduce cost, increase access, and grow farm businesses for producers and their cooperatives.

An example, NOW Wireless, the Missouri Farmers Union in response to the demand for affordable modern telecommunications access for farmers and rural residents living in remote areas, helped established USA Broadband. USAB has partnered with subscriber-based cooperatives and develops successful networks that are making this access possible.

Today, USAB is a premiere provider of high-speed wireless broadband Internet, voice communication, and video services to rural communities. The community maintains its focus on providing a superior broadband product back with exceptional customer care. In fact, about a year ago USAB partnered with the Eastern Illinois Electric Co-op, a member-owned cooperative to develop a broadband Internet network that focuses on providing a wide range of Internet broadband services to rural residents within the cooperative's 10-county service area. This includes 6,000 square miles and 240,000 homes and businesses. We also plan to utilize our credit union that we have for rural residents and farmers to help finance the services and equipment, especially for our more disadvantaged residents.

Internet is a necessary tool for farmers and ranchers who will be at an economic and competitive disadvantage if unable to use the same high-speed Internet connections that are available to other small businesses around the country and around the world, as far as that goes. USDA encourages farmers and ranchers to rely on the Internet to check weather, market, crop reports and file applications for federal programs. However, for many rural producers, the reliance on the Internet cannot be a reality.

Given the current economic climate, it is imperative that producers devote as much time as possible to marketing their products, exploring new markets. The ability to conduct financial transactions on-line would save individual producers hours of administrative work and translate its tremendous financial incentives at the farm level.

NFU has a program called e-cooperative.com. It is the world's first innovative portal to directly locate and buy quality food products plus other goods and services on-line from hundreds of U.S. agricultural producers and their co-ops in rural America. This basically has allowed us to build authentic relationships between farmers and consumers. In fact, I'm president of a co-operative that producers and processes natural pork and we market it throughout the country and the world, in fact. A lot of our pork goes into New York.

And so as we go forth, this is basically the new renaissance in agriculture and it's so dependent upon modern telecommunications. By eliminating the digital divide and providing more rural areas with high-speed Internet access, we can help these producers and these new producers market and sell their quality products and educate consumers about the value of family farmers and ranchers.

There's real challenges we heard a lot about, some of the challenges about the reluctance of these providers to come into more remote and under served areas. It's been a challenge to secure financing from providers because we're kind of—we're often are at an awkward size loan fund, too small or too large for some of them.

We also have challenges that we're concerned with limited competition in the rural markets. Some of the solutions might include that we've heard before the possibilities of providing tax credits and other incentives to inventors that want to invest into the more remote areas, as well as the possibility of allowing a 10 percent match on the Rural Utilities Service Loan Program, rather than the 20 that's necessary. It's also possible and suggested that maybe the FCC has the ability to reallocate frequencies that will become available in television's transition to digital. We propose that successful applicants to the rural broadband initiative program would be granted the license frequency resulting in improved equity for rural broadband service providers.

Better broadband means a better place to live in rural areas and we appreciate the interest and we really believe that rural broadband is the key to rural revitalization.

Thank you very much.

[The prepared statement of Mr. Kremer may be found in the Appendix on page 80.]

Chairman Shuler. Thank you, Mr. Kremer, and congratulations to farmers to get them back into economic structure. They need all the help that we possibly can provide for them and I thank you for your hard work and dedication to them.

At this time the chair will recognize Brent Christensen. Mr. Christensen is the vice president and general manager of Christensen Communications Company, an independent, local, exchange telecommunications carrier in Madelia, Minnesota. Founded in 1903, Christensen Communications Company provides local, long distance and cellular telephone services in addition to dial-up and high-speed Internet services. Mr. Christensen also serves as a chairman of the Legislative Policy Committee for the organization for promotion and the advancement of small telecommunications companies.

Mr. Christensen, thank you for being here today and we look forward to hearing your testimony.

Statement of Brent J. Christensen, Vice President and General Manager, Christensen Communications Company, Madelia, Minnesota

Mr. Christensen. As he said, my name is Brent Christensen. I work for a telecommunications company in Madelia, Minnesota, population of about 2300. It would be easier to tell you that we are a telephone company, but quite frankly that's no longer an accurate description.

I also have the privilege of serving as the chairman of the Legislative Policy Committee for OPASTCO. We are very integrated in our community. We employ six people, not counting my parents and me, and all but one of our employees reside in the community. We encourage our staff to be active in the community. I currently serve on the Madelia Public School Board and am vice president of the Chamber of Commerce. I also previously served as the Mayor of Madelia, and president of the Madelia Development Corporation.

I'm here today to talk about broadband's impact on rural communities and Madelia, in particular. We started providing broadband service in 2000. We didn't start by putting a business plan together and figuring out how much money we could make. We started offering DSL because it was important to the economic survival of our community. We entered into the DSL business because Marv Davis needed it. Marv and his son, Will, own Davis Sales and Service, a local Polaris dealer. We had been offering dial-up Internet service for a few years, as was the competitor. They told me that Polaris had changed the way that they sold their snow mobiles, watercraft and ATVs. Warranties were now issued over the Internet. When a customer came in to buy a snow mobile, the Davises would fill out the customer information on-line and print off a warranty application. Once the customer had signed the document, the Davises would scan the document and transmit it back to Polaris over the Internet. The problem was that dial-up was too slow for this process and their dial-up connection would frequently time out and they would have to start over. This was frustrating for the Davises and their customers. In the end, if we didn't solve the problem, the Davises would sell fewer Polarises and it would severely impact their business.

I did some research on different solutions that would work with our network. We bought some equipment and we got DSL service to the Davises. The entire process took about 20 days. We didn't do a business case first. We didn't go through any corporate bureaucracy. We just got a new service to a customer that needed it.

When I was in high school, I worked for the telephone company as summer help. My grandfather was president of the company at the time and I remember the two of us walking back to the office one day and him telling me how important the telephone company was to the community and how we had a responsibility to provide the best service possible. Back then it meant providing quality, reliable telephone service. Today, it means so much more. Today, we have to provide state-of-the-art communications for the survival of our small town.

Madelia is a lot like other towns our size and in many ways like the communications industry itself. We are in competition with other communities in our area. We are in competition for industry and people. As a community, we have to leverage our assets to develop our economy. Communications is one of those assets. Because of our communications infrastructure, we can market our town to telecommuters, small businesses, and others who do not depend on a specific location to conduct their business. A good example of this is the House of Print. They're a local printing company that was started in the 1960s by a company that owned two daily newspapers in towns 20 miles from Madelia to the north and south. Both papers needed to replace their printing facilities and instead of each buying new presses, they built a new printing operation in Madelia which is halfway between the two. Today, the House of Print prints for over a 100 daily and weekly newspapers. The House of Print was our third DSL customer. Our high-speed Internet allowed them to expand their customer base and increased their business. They have literally brought in millions of dollars of new business because of their high-speed Internet connection.

The House of Print is no longer geographically limited. Today, they can bid on printing jobs on-line, allow the customer to upload data, proof the job and mail the finished product directly from their facility. The House of Print has expanded significantly as a direct result of the Internet, and they have added or upgraded their printing presses and expanded their building facilities.

As a small rural company, we face many challenges providing state-of-the-art communications. We have to provide all of the same services as larger companies and this gives us a good understanding of our customers. An example of this is Farmers State Bank in Madelia. They are a locally-owned independent bank. They compete against the Madelia branch office of a much larger bank. Our high-speed Internet connection has allowed Farmers State Bank to offer a full line of Internet banking services. These services have kept Farmers State Bank competitive with other banks in our area.

Companies like Christensen Communications look to Congress for leadership on issues and programs that give us the opportunity to thrive and in turn keep our customers and community thriving. We ask Congress to continue to support a strong and viable Universal Service Fund. The USF is the most important federal program for our continued success. Congress and the Federal Communications Commission needs to support the reform of the inter-carrier compensation regime by implementing the Missoula Plan which was developed by a broad cross-section of the telecommunications industry. And Congress needs to support programs like the Agriculture Department's Rural Utility Service and the Small Business Administration that helps small businesses like mine.

I thank you for the opportunity to testify today and I would be happy to answer any questions you may have.

[The prepared statement of Mr. Christensen may be found in the Appendix on page 85.]

ChairmanSHULER. Thank you, Mr. Christensen. The chair will now recognize Ranking Member, Mr. Fortenberry, for the introduction of our next two witnesses.

Mr. Fortenberry. Thank you, Mr. Chairman. I'd like to introduce Mr. Brian Mefford, good morning. Brian Mefford serves as the president and CEO of Connect Kentucky, where he is responsible for leading the successful implementation of Kentucky's prescription for innovation, a comprehensive plan to accelerate technology availability, literacy, and use. During Mr. Mefford's tenure at Connect Kentucky, the organization has evolved from a research-focused business with a staff of five, to a technology-implementation business with 35 staff members working statewide. Prior to this role, Mr. Mefford served as Kentucky Chief of Staff to Kentucky Commerce Secretary, Jim Host.

Welcome, Mr. Mefford, please give us your testimony.

Statement of Brian Mefford, President and CEO, Connect Kentucky, Bowling Green, Kentucky

Mr. Mefford. Thank you, Ranking Member Fortenberry, Chairman Shuler, and Members of the Committee, thank you for the invitation to be here with you today. I appreciate it tremendously and I appreciate the opportunity to discuss this important issue.

I am also the CEO of Connected Nation which functions as the parent company of Connect Kentucky which was, in essence, our demonstration project. It's the Kentucky story that I am here to discuss with you today.

I first want to open by briefly talking about a handful of stories that illustrate the entrepreneurial environment that exists today in Kentucky. First, I'll mention Cameron Cohlsen, who like a lot of Kentucky natives, after graduating from college found opportunities to work outside of Kentucky, found himself in Boston working for a creative services firm. And one day he looked around and said you know, it's about time I move back to be with family in Kentucky. And he said there's no reason I can't do what I do here in Boston from home in Kentucky. And so he went back to his family farm. He grew up in a family of farmers and he went back and he looked at about a 100-acre lot that had been—where his family had raised barley and tobacco for years. And he said I can envision my new creative services building right in the middle of this what was now a pasture. And so he built that company and indeed did everything that he was doing in Boston, has grown that company. Last year, won a competitive bid to produce all the creative paraphernalia and programs for the Academy Awards. And Cameron reported that having a broadband connection on that—on his family farm in the middle of Kentucky was just like being right down the hall from those folks in California as they were trading photographs and things to prepare for the Academy Awards.

And there's Global Data Tech, a company saw an abandoned mine in an eastern Kentucky county, Appalachian Regional Commission, a distressed county, and saw an opportunity to create a business there doing underground, subterranean data recovery, disaster recovery, and data backup. And so that company is now working and creating jobs that are high paying, mostly above $60,000, $70,000 jobs in this county and helping that community to flourish.

Then there's Jared Fugate, an individual who I received a call from a couple of years ago and he said you know, I'm having to move with my family out into a more rural part of my county, again in eastern Kentucky. He said I have a three-year-old business. We do tech support and web development and he said there's not broadband service out where I'm going. I really need it or my business is going to fail. Long story short, after I talked to him more, I said tell me more about your business and how it started, those types of things. He said well, I'm 17 years old. And I've been building websites for the businesses in the community since I was 14 and it's evolved and it's a company that I plan to run through college and come back here and maintain after I graduate.

Farmers were mentioned earlier. We have poultry farmers who are managing temperatures of poultry houses remotely from all places, parts of Kentucky. We have folks who are managing farms, not just in tracking soil quality and weather, not just in places across Kentucky, but now in places around the world where folks— farmers are able to invest in land and other places in the world, they're able to do that type of monitoring from their homes in Kentucky.

I wish I could that environment, that type of environment has existed always in Kentucky, but it just hasn't. In fact, four years ago, Kentucky faced the same challenges that are all too common in rural parts of the country in rural communities everywhere. The Commonwealth was struggling to use technology-centered solutions to address traditional challenges related to economic development, health care, education, and delivering government services. On the economic development front, jobs and manufacturing, farming, and mining were leaving at an alarming pace. The indicators of Kentucky's technology troubles were not hard to identify. Kentucky consistently ranked low, in fact, at the bottom of the barrel among states in terms of broadband availability, broadband usage, the number of high tech companies at work in the Commonwealth, and further, college graduates were leaving in droves, creating what we all know too well as the brain drain effect.

So as we surveyed the landscape for answers, the reality of the situation was certainly troubling and we realize that the foundation of broadband infrastructure was not adequate for creating solutions that could address the challenges of a new day, not adequate to provide widespread access to telemedicine, distance learning, and e-government, and not adequate for growing and attracting entrepreneurs and industry, not adequate for providing more opportunities to our farm families and communities where our children were leaving the rural roots never to return.

And so Connect Kentucky set out to identify the root cause that had resulted in the lackluster technology picture. And it was clear that the inadequacy of Kentucky's broadband infrastructure could be traced to much of the state's inability to compete in so many areas important to the knowledge-based economy. And so broadband infrastructure had been built into those more populous areas as several folks had mentioned before, but it was those rural areas, less metropolitan areas that were under served. And so the lack of service not only created the well termed digital divide for rural residents, but it also made it impossible to create state-wide policies and initiatives that can make the entire Commonwealth more competitive.

Further, we discovered that broadband availability was only half of the challenge. It was broadband usage that represented the other part of this challenge that had to be overcome and we realized that any comprehensive strategy had to address both sides of that equation. So next we identified the barriers that were inhibiting broadband availability and use. In terms of availability, there were a series of issues that needed to be addressed. First, very little data existed to allow us to identify the true extent of the broadband gaps in Kentucky. Providers didn't know. Policy makers didn't know. And communities themselves didn't know.

Chairman Shuler. We'll have you finish up more of your testimony during some of the questions, if that's okay.

Mr. Mefford. Yes, sir. Would you like me to close?

Chairman Shuler. Yes. That's a nice way of saying it.

Mr. Mefford. Well, the results which I can address during questions have been that we developed maps, as was mentioned earlier, that identified the broadband gaps. We worked with providers to address market intelligence, to provide market intelligence that lowered the cost of entry into our rural markets, and we aggregated demand and helped create demand at the local level. And bottom line is we used a public/private partnership to lower those costs of entry to incent market effects, to make Kentucky a more attractive environment for broadband development and use.

Thank you, Mr. Chairman.

[The prepared statement of Mr. Mefford may be found in the Appendix on page 91.]

Mr. Fortenberry. Now I'd like to introduce Mr. William Deere. Mr. Deere serves as vice president of Government Affairs for U.S. Telecom, a trade association representing 1200 member companies offering a wide range of services including local exchange, long distance, wireless Internet and cable television services. Before joining U.S. Telecom, Mr. Deere served as Deputy Assistant Secretary in the Bureau of Legislative Affairs at the U.S. Department of State where he oversaw the Department's legislative initiatives. Mr. Deere also worked on Capitol Hill as a House Appropriations Committee Staff Member for Representative Jim Lightfoot.

We welcome your testimony.

STATEMENT OF WILLIAM R. DEERE, VICE PRESIDENT, UNITED STATES TELECOM ASSOCIATION, WASHINGTON, D.C.

Mr. Deere. Chairman Shuler, Ranking Member Fortenberry, Members of the Subcommittee, thank you for this invitation to be here today. U.S. Telecom is the nation's oldest and largest association representing rural telecom providers. The vast majority of our member companies are rural companies. They are small businesses serving small communities.

We were pleased to see affordable broadband access for all Americans as a component of Speaker Pelosi's innovation agenda. In addition, the Senate Republican High-Tech Task Force is calling for policies that promote widespread broadband deployment and use of broadband technology. Broadband deployment and adoption should be nonpartisan objectives and we believe the Congress, the FCC, and the RUS have vital roles to play in advancing these goals for rural America.

U.S. Telecom and our member companies are committed to furthering rural broadband deployment and believe the Congress can advance a number of initiatives that promote this goal. First, we must ensure a sustainable future for universal service. In the House, Representatives Boucher and Terry have recently reintroduced universal service reform legislation they first proposed last year. We appreciate the Congressmen's dedication to finding a sustainable long-term solution and we hope Members of this Committee will encourage consideration of such reform legislation this year.

In addition, Congress can promote broadband deployment by, among other things, permanently extending the Internet Tax Moratorium. U.S. Telecom supports H.R. 743, bipartisan legislation to extend the Internet Tax Moratorium that was introduced by Representatives Eshoo and Goodlatte. I encourage all Members of the Committee to consider co-sponsoring this legislation. I know that Ms. Musgrave, and Mr. Fortenberry, you are already co-sponsors of this bill, and urge the House to take up this important legislation before the moratorium expires in November.

Finally, in its relatively brief history, the RUS broadband loan program has achieved some successes. But we believe with modest changes largely based on the successful RUS telephony program, it could accomplish even more. We were honored to appear last week before your Subcommittee, Ms. Musgrave, of which Mr. Fortenberry is a Member, to offer some suggestions for the upcoming Farm Bill. We believe the primary weakness of the current program is that it does too little for areas with no access to broadband. And while the current practice of offering cost of money loans makes projects financially viable in some areas, other high-cost areas will require low-cost loans or a combination of loans and grants to make a costly infrastructure bill feasible.We also believe steps should be taken to expand the number of companies eligible for broadband loans and that steps be taken to improve processing of loan applications at USDA.

Finally, we second Mr. Mefford's testimony. By the end of 2007, Kentucky will go from having one of the lowest broadband subscription rates to having broadband available to 100 percent of its households. That's impressive progress. And we think Congress might look to Connect Kentucky as a model for what works.

Modernization of the nation's communications infrastructure will seed economic growth and expand opportunities. Nowhere in the nation do these advances hold more potential than in rural America. We thank you for your invitation to appear today. U.S. Telecom and its member companies look forward to working with the Subcommittee and this Congress to achieve our shared objective of making broadband as ubiquitous today as electricity, water, and telephone service.

[The prepared statement of Mr. Deere may be found in the Appendix on page 104.]

Chairman Shuler. Thank you. I'd like to thank all of you for your testimony.

Mr. Mefford, can you describe some of the types of public/private partnerships that Connect Kentucky actually really benefited, I mean outside of the—and I want to commend the 17-year-old young man for his outstanding work in entrepreneurship.

Mr. Mefford. Yes, sir. Mr. Chairman. First, it had to do with the mapping initiative that we undertook. We realized that our folks in government needed batter data. Our policy makers needed better data. We realized that our providers were willing to some degree to provide, to give that data and they actually wanted more data about unserved areas. They wanted to cooperate with one another to figure out where there unserved areas where.

And so we put together this initiative to map, to create this broadband inventory where every single provider in the state provided their data to us of where their served areas were. And so in your packets, I believe in front of you, perhaps, there are examples of our maps where you can see all across Kentucky, all the different types of service that's available.

But more importantly, we can then invert that data, that map, and focus on those unserved areas and we can start drilling down into the data overlays that then help providers identify low-hanging fruit immediately, providing household density data, for example. And then we can overlay data household survey data where we ask households would you adopt, if broadband was available in your household? What price points are you comfortable with, that type of information. And so we can really provide through that public/private partnership market intelligence that allows providers to move more quickly into unserved areas. I'll mention on the demand side too, at the same time we're working with communities.

We have 120 counties in Kentucky. We have created what we call E-community Leadership Teams in every county. Those are teams that represent a cross section of the community. We have farmers. We have local government officials. We have business and industry, educators, health care officials, on and on, who come to create a strategic plan for how to use technology as it's coming into the community, how to better use it and plan for. In effect, as Commissioner Adelstein said earlier, we're creating demand and aggregating demand at the same time. So the two processes work hand in hand and allow providers to see more opportunity and have a better business model for moving into these unserved areas.

Chairman Shuler. Mr. Kremer, can you give some examples or some ways that our farmers can actually utilize the access to broadband to really strengthen their farming community? Some of the farmers in my area, they probably say they don't know how to turn a computer on. So how can we actually incentivize them to truly become more computer literate then and obviously extend it through, their business through the Internet?

Mr. Kremer. I have some good examples. I'm also a former agricultural educator that started the Young Farmer Programs and it's interesting that during the early '80s when farmers were completing scared to death of computers, I actually took the computers, put them on the kitchen table and those dual floppy drive Apple computers and taught them to get over the fear and use those computers. I think we've got to do the same thing with broadband. And I truly believe that for small and moderate size farmers, this is probably the last hurrah that we have is to build those authentic relationships with consumers. And this is how they can do it.

I mean even people like my colleagues in North Dakota, for instance, say that Missouri is more advanced because we have population centers. Well, the Internet, high-speed Internet makes the world smaller in building relationships. We have, for instance, in our organization, have developed artisan cheese plants, for instance, made from goats in Southern Missouri, Goats R Us, basically, and have taken this artisan cheese and basically made it very special and famous and have sold it throughout the country. Have done the same thing with exclusive type of heritage vegetables in Southeast Missouri. We've done the same thing with beef and pork and dairy products, and even wood products. It's basically being able to have access to that type of technology.

And what this also does it brings in and encourages and retains our young people with the brilliant, vibrant minds and those people graduate from college and can come back to the very rural area and make that connection. And so it's extremely important.

One last thing also and something I want to point out is that for instance, even our USDA and the Farm Service Agencies have been attempting to close some of the offices down which are very vital to our rural communities right now because of lack of computer access. Their point is that farmers should be more Internet savvy and access those programs on those farms via the Internet line. Well, we can kind of accommodate, but until we have more service in the rural areas we will continue to fight a lot of those closings, but this would help solve some of those problems and save the government and overall consumers some money.

Chairman Shuler. Mr. Stephens, can you tell me about some of the barriers that you are facing in expanding the broadband in your area?

Mr. Stephens. Some of the barriers that we are facing basically are, as I described, were somewhere terrain and others we need just the basic support of placing the infrastructure there. We do not exactly have all of the public private support that we need. Because if we had, we would have never formed Balsam West FiberNET to begin with. It would have been at the hands of the public or the private sectors.

The public sectors with the local government: (1) Does not have the expertise nor do they have the funds, or do they have access to the funds to build the infrastructure. The low level population in the area, the scattered areas, the isolated areas; all of this came into be. As you all know of the mountains of western North Carolina and some north Georgia and eastern Tennessee, that's just to name a few things.

Otherwise, we do have a few things leading into the area, what has been placed into some of our rural mountains areas, what the government has provided and has placed there has not be effective. And to those areas I think sometimes coming at some inappropriate oversight to those resources that have been promised have not lived up to the promises to the areas.

Otherwise, some of the barriers, I think we were describing earlier from some of the other panelists, is that it is basic resources for technical assistance to training our residents on how to use technology. We have a great highway. We have great resources to put those people down that highway of information. But we need to have some skilled people to drive the Cadillacs down the highway, too.

Chairman Shuler. Are you finding problems or issues actually connecting with some of the more of the Federal funded programs?

Mr. Stephens. We are. We did have one resource that came to western North Carolina that was known as the ERC, the Educational Research Consortium that was set up in western North Carolina that based itself in Asheville. And the problem was to deliver high speed Internet, to deliver broadband data transport to the areas to connect up with our school systems, do these things. And we are not yet seeing that happen. That has not occurred. So those are the things that are happening.

Otherwise, we are falling through the cracks of eligibility in other programs such as RUS. We have too much population. Again, as Commissioner Adelstein told you earlier, they look at a map and say you have DSL in this area when one person has that. We have broadband. And that is just not the case. I think those models of problems that you are seeing, that you are hearing from other panelists speaking or Commissioner Adelstein earlier, I think that mirrors what we have in western North Carolina.

Chairman Shuler. The Chair will now recognize Ranking Member Mr. Fortenberry for his questions.

Mr. Fortenberry. Well, thank you all again for appearing today, and I appreciate your insights for most of you, particularly your innovations in being entrepreneurs and the effective and aggressive utilization of broadband services in your particular communities. It is very impressive what you all have done.

I noted that your comment, Mr. Shields, about the potential to enhance telecommuting, which is essential for the well-being of rural communities. We have talked about potential value for small business, the potential value for education and health care delivery. But the potential value of allowing a person to telecommute is very substantial as well, and it is an important point.

Now you can go on line and put your order in for a McDonald's hamburger, is that what you meant in North Dakota?

Mr. Shields. Yes, I was actually at a farm in North Dakota yesterday with one of our employees showing Toyota people how it worked. And McDonald's is working to change the taking of orders.

When you go to a McDonald's drive-through, you drive up to a speaker post, you talk to somebody. And they want to move about 10,000 jobs out into the rural communities to be the ones who are on the other side of that speaker post instead of having somebody doing that in the restaurant. And we have been very pleased with the work and the quality. And it is one of many things that will be done over the next decade to move what were call center areas like in Lincoln, and I actually was a kid in Lincoln, where you have got a lot of the IT services; Lincoln, Omaha, Sioux City, Sioux Falls, Fargo, Grand Forks where we are going to now move out into these communities. Because we do not need the walls anymore. We do not need to have people sitting in a big building. We can have them work wherever they want.

And this woman whose farm we were at was able to work for the first time in 20 years. When she got married and was out there in the farm, there was nothing she could do out there. The family was able to add a brand new add-on that she was proud to show us to her house because of the earnings that she was able to get from that job.

Mr. Fortenberry. Well, it is an outstanding example that concretizes the reality of how this potential technology can achieve so much social good in addition to advancing the movement of products and services.

Mr. Kremer, you very well point out how the small and mid-size farmer is going to increasingly depend on this connection direct to the customer.

One of the most successful farmers I have in my District was a struggling commodities farmer who changed his business model and relies heavily on the Internet sales of a specialized hay product now for pets. So it is a great point.

Mr. Christensen, I particularly picked up on the fact of your leadership. All of this is important. Advancing the technology is important, but having leadership in localized community, I think you mentioned your—what did you call it, Mr. Mefford? Your e-leadership council?

Mr. Mefford. E-community leadership.

Mr. Fortenberry. Is a great point as well that I think we can all learn from.

I would like to further unpack some of your efforts in Kentucky. I think one of the considerations that may come out of this hearing because we have discussed it, Mr. Deere, you have touched on it as well, is the need for a clearer understanding perhaps through mapping or some other mechanism, but I think mapping is the core issue here of where there is true broadband availability, where it is underserved, where it is lacking in service. Who is Connected Nation? Was this founded in Kentucky? Are you the primary principal? What is the main objective of it? And is using a smaller footprint such as a state to provide this mapping or to provide as an entity in a state to potential provide the capacity for a more aggressive mapping system could then be duplicated in 50 States, is that a better platform than perhaps we mandating it at the Federal level as a joint project, perhaps, of the FCC and the Rural Utility Service? I think that is one of the core points that both you gentlemen made as a key component to ensuring that limited funds are targeted precisely or the private market actually has better information to develop its services.

So two points. Explain better your mission, how it was conceived, what your long term vision is? I am curious. Obviously, you are doing tremendous things. And then touch upon the mapping issue from a state perspective how that can potentially be used nationwide?

Mr. Mefford. Yes, sir. I appreciate the question and the compliments as well.

We envision that if we could demonstrate the success of this model as it was set up in Kentucky, we envision that it would be highly transferrable to other States. I believe it is Commission Adelstein that said in the past, and I say it often, there is not anything that is on the level of rocket science about what we have done. I mean, it is really a lot of entrepreneurial type ingenuity. And I think Commissioner Adelstein said Kentucky elbow grease is how it got done. But given that fact, it is highly transferrable. And so we had a lot of inbound calls from other states and established Connected Nation as the national nonprofit means by which we could transfer that model to other States, and also work in cases with the Federal Government as well.

Mr. Fortenberry. And how are you funded? What are your funding sources? Is it a consortium of interested private sector parties? Do you receive grants from the State? Federal money as well?

Mr. Mefford. Connected Nation is working with States. And so we primarily are receiving State funding as Connect Kentucky has. Connect Kentucky is both public/private. And so we have some contributions from participants from the private sector, but it has primarily been State monies.

Public/private is ideal so that you do have that type of engagement from the private sector community as well as from State resources, State and Federal resources.

Mr. Fortenberry. What I hear you saying about the mapping process is that it is half science, half art. You are somehow able to achieve proprietary data that, as Commissioner Adelstein was saying, is a problem on the Federal level. You have been able to do that, and I assume that is through relationships? Melding the art and science of getting this end task completed?

Mr. Mefford. They are relations that, again, transfer to the national level as well. And so those relationships are in place now to be able to do similar things that we have done in Kentucky nationally.

Providers are attracted to an independent third party as a place to aggregate this data. They want it protected, to a certain degree. And so we are able to sign nondisclosure agreements that protects the data down to a certain level of granularity.

And so on our maps you can see where the service is and what is provided, but it does not detail who is providing that service, in most case. Or in all cases, rather.

And so there is that desire from the provider community to have that independence, that independent third party.

Now to your question of do we go at this State-by-State in terms of the mapping or do we do it at the Federal level? I do not think they are mutually exclusive. In fact, the model in Kentucky bears that out that both is likely a better answer. In other words, the Congress could empower the FCC and enable the FCC to begin such a program, hopefully in partnership with an independent entity where it is a data clearinghouse that is established as an independent entity. And so you have a federal effort. But then as we did in Kentucky as we moved community-by-community, it is sort of a verification that takes place. And that is a bit of a slower process, but it is going into the States to take the data that has been aggregated federally and fact check and say is this right on a very granular community level.

So I believe that it could be both, in answer to your question. That it could be a Federal effort and a State-by-State effort. In fact, there is a Senate bill that has been filed, 11-90 by Senator Durbin that sets up a State-by-State approach. And, again, I do not think it is mutually exclusive to have a bill that establishes the Federal approach as well.

Mr. Fortenberry. But again in terms of the fundamental platform by which your successful efforts have been achieved, it is that set of data? I have heard you say two things. That has been critical and, obviously, leadership and desire down to the smallest local entity has been critical. But I do not want to set up a framework here where there could be a competing framework by which we can leverage our resources to better provide broadband throughout the country. But it seems to me that a lot of this testimony is pointing to that pillar of just better understanding of market data that is out there that can be used to create momentum synergy, better synergy in the market as well as more targeting of our limited Federal funds. Is that a correct assumption?

Mr. Mefford.That is right. Absolutely. I think it has to be the tip of the spear. We do not know where we are today. We just do not. And that map, that national map really has to be created.

Mr. Fortenberry. And I appreciate your comments about a potential hybrid solution versus an either/or; Federal or more Stateby-State projects. So thank you.

Would you care to comment on that, sir?

Mr. Deere. I would second Mr. Mefford's comments. And when we appeared before you last week to discuss to discuss the RUS program, there is not a single magic bullet that solves the problem. It is making sure the funds go to underserved areas. It makes the loans more creative to go into areas where there is a truly bad business case to be made for issuing the loan. And it is working through programs like Connect Kentucky to make sure that we know where we need to go.

As you heard in the first panel from Commissioner Adelstein, there will be issues that have to be worked out at the Federal level when we start looking at a national mapping plan. In fact, I believe as early as next week the Telecom Subcommittee, Mr. Markey's Subcommittee, could be taking up the issue. But what we were trying to point out today and last week was we have a model that works. And this is a bipartisan issues. And this is a Congress that wants to get things done. And I think we should move out on something like Connect Kentucky.

Mr. Fortenberry. Okay. Thank you.

Thank you, Mr. Chair.

Chairman Shuler. The Chair will now recognize the Gentlewoman from New York, Ms. Clarke.

Ms. Clarke. Thank you very much, Mr. Chairman.

And I have to tell you gentlemen, this has been very, very, very groundbreaking for me hearing and your testimonies just have been fascinating. What you have been able to do through ingenuity and really just out of necessity in utilizing the technologies that are available to us to really make our world a little bit smaller and make our communities much more connected is what is needed in the 21st Century. And there is no other way for us to go. And you are the trailblazers. And I have a feeling that we will be looking back at your gentlemen in generations to come as sort of like the Henry Fords of what has led our communities and in our nation into its strength and really coming into its own as a first world nation.

Let me direct my questions Mr. Mefford. Because Connect Kentucky is truly fascinating. And, again, groundbreaking. And it has demonstrated a willingness to improve the quality of life for low income communities, spurring and cultivating local entrepreneurship. Connect Kentucky has made strides using technology to improve health care, education and community development.

Can you just briefly tell me some more about your challenge as a CEO, challenges with regard to cost factors? And do you believe that you could replicate this model in urban environments that have a similar struggle?

Mr. Mefford. Yes, ma'am. Thank you for the question.

As I mentioned, the model is highly replicable, can be transferred to any State. And all States have these challenges, because all States have either urban gaps or rural gaps, the broadband digital divides. And so we have heard from nearly every State that wants to replicate all or part of what we are doing in Kentucky.

The primary barrier, as you mentioned, is funding. And so that is why we started working here on the Hill to discuss the importance of creating some enabling legislation that allows States to develop programs that are empowered at the Federal level but are localized in their design. That is an important element of being able to transfer this across the country.

And to your question of does it apply in urban areas. Absolutely. And so in each of our communities where we can identify an urban digital divide, we have been addressing those just as we do in the rural areas.

We developed a program. In fact, their conventional wisdom had said in the past that folks did not adopt broadband because it was just too expensive. But we found that the primary reason that folks were not adopting broadband, particularly in our urban areas, was that they did not own a computer. They did not own a computer and it was too expensive or the second was the answer you said, it was understanding that it exists or knowing that it was available.

And so we launched program and knowing that granular kind of data, that market intelligence, that we launched a program called No Child Left Offline. And so that uses State refurbished computers and adds Microsoft software and CA software, security software and puts those computers in the households of underprivileged children. And so it is through efforts like that where we can then start impacting the adopting of broadband.

We have had 73 percent increase in households actually subscribing to broadband over the last two years.

Ms. Clarke. I would like to direct a question to Mr. Stephens. I recently read an article in the Asheville Citizen-Times, which is published in our Chairman's wonderful State of North Carolina, and it is that your company has formed a partnership amongst rural school with Drake Enterprises and the Eastern Band of the Cherokees to bring high school Internet access to 45 rural schools. Many experts see high speed Internet access as one way to help reduce poverty and close the digital divide.

With the improved access to high speed Internet capabilities children would enjoy enhanced educational opportunities and their parents could learn their skills necessary to thrive in an increasingly computer-based economy.

How successful has this partnership been and do you have any plans to pilot this model in urban environments such as New York?

Mr. Stephens. Well, Ms. Clarke, I believe that there is really no difference between a green covered mountainside and a skyscraper, really. It's just an obstacle in the way. And, of course, we in western North Carolina take great pride in those mountains that give us strength.

But, yes, this model has been very successful because the high cost of services provided to our schools, provided to our businesses, to individual was outrageous. We were paying eight to ten, twelve times the rate that metro carrier rates would pay in other urban areas; Atlanta, Knoxville, Charlotte, those areas.

One of the things that we had to do was to be able to bring that access to the schools. Furthermore, we had to give them the opportunity to choose their provider. To choose the content that suited their needs for their curricula, to suit their needs for their culture.

The Eastern Band of Cherokee Indians is working to build an entire new school system, if you have seen any kind of congressional in the past couple of years. We have been working to build an entire new school system. The curricula there is going to be a little bit different.

So for all types of different reasons, yes, this is a successful model. It is taking the access to the schools. They own the fiber optics. They own the content. Basically they are in charge of their own destinies. And right now the construction is beginning on those laterals and on those star networks for the school systems. Unfortunately, as some of the other panelists have pointed out, funding is a major issue here. In order to: (1) Complete all of the connections to all the schools. Because our dream here is to have "No Child Left Behind," no school left behind to be able to inspire any kind of creativity for the entrepreneurship to breed home grown businesses, to be able to allow them to stay in the area. There has been a serious out migration of our best and brightest. I mean, it is great that they are able to get the education that they can, go to the University of North Carolina, go to Harvard or wherever they can go, but they seldom come back because the opportunities do not exist.

So is it successful? Yes. Because we are putting that access there. They are control of that destiny. So for that. We are also cutting down the costs. It is a wonderful model. And, yes, we hope that it is something that can be replicated throughout the area.

We are starting in our State. And we hope that this will embers the big brush fire that burns across the State.

Ms. Clarke. Mr. Chairman, I just want to make a comment to Mr. Kremer, and maybe to all of you.

One of the things that we are looking at in urban areas with respect to the Ag bill is the use of food stamps to be able to bring nutritional foods to communities that do not have access to them. You would be surprised at how many urban areas where green groceries are not readily available, where quality or now we are moving into organic kinds of foods that lead to proper nutrition of Americans are not available. And just in listening to what you have said, one of the things that has been a challenge is how do we get or create an avenue for those who are less able financially to shop in the markets but have to utilize Government subsidy of some sort. To work with the small farmer in terms of co-oping or things of that nature and being able to access those types of nutritional things.

Have you had any conversations or have you had any thinking around that? Because I think that is one connection. You already talked about the pork. I am going to be looking for that. But, you know, other items that we can connect to other people in other parts of the country to really begin to address those issues in our communities?

Mr. Kremer. Yes, I have thought about it and I share your same concerns, and it is something that is part of our vision and our mission. And that is to provide this type of wholesome food that has got this story behind it and authenticity and make it affordable and accessible to all people.

And when you talk about, for instance, rural broadband and connecting these smaller entrepreneurs and what I would call collective entrepreneurs, you know we have these little co-ops, for instance, in remote southern Missouri that our prices are not the highest. They are kind of like in maybe in the upper one-third share of what conventional pricing would be. And so, but it has got greater value.

For instance, you may think this is an oxymoron that we have a healthy hot dog. That has no additives, that is not filler, no chemicals and that school systems want this. And I do believe that it is in the public's interest to somehow subsidize school systems or whatever, and the food stamp program as well, to allow disadvantaged people to access this as well. And I think it could be a very affordable program that is in the best public's interest.

Chairman Shuler. Mr. Christensen, you had spoken earlier about ways you have been able to maintain businesses at home instead of having to go out of your District or maybe into more urban areas. Now what are some of the ways you are highlighting in your community, ways to actually say here is how broadband has worked, here is how it has worked for us, here is how it can work for you?

Mr. Christensen. Well one of the ways is coming out here and talking about it. A lot of it we are doing through the Chamber of Commerce. We are doing through our community education program.

Some of the other people talking earlier talked about getting people trained on using the Internet. We are a highly agriculture community. So our community ed program puts on classes on Microsoft Excel and Outlook and things like that for senior citizens or people who do not have a lot of experience with it.

A good example of one of these programs is as a result of one of our community ed programs a guy that retired from the local hardware store started an eBay business. And he sells tractor manuals online on eBay and works out of his home four miles from town, and has been able to stay on his farm.

So that kind of word spreads around. And he has now come back and he teaches class to his peers. And that is really what we are doing.

Chairman Shuler. Well, I would like to thank everyone for their testimony today. And just extend a special thank you for what you are doing with our rural America and in the urban areas to really expand our businesses, small businesses in particular, and what you have been able to accomplish. Continue your hard work, your dedication. And I will look forward to working with my colleagues on the issues that have been raised today for us to create the proper legislation that will actually work for the small business in America.

At this time the hearing is adjourned.

[Whereupon, the Subcommittee was adjourned at 12:01 p.m.]

STATEMENT
of the
Honorable Heath Shuler, Chair
Subcommittee on Rural and Urban Entrepreneurship
Hearing on Maximizing the Value of Broadband Services to Rural Communities
Wednesday, May 9, 2007

I am pleased to call to order this hearing. This is the Subcommittee on Rural and Urban Entrepreneurship's first hearing of this 110th Congress. I would like to welcome the Members of the Subcommittee, the distinguished witnesses, and our guests.

Rural America is home to many different kinds of small businesses. In Western North Carolina, there are thriving high tech businesses, small manufacturers and family farms. What all of these businesses have in common is a need to stay connected with their customers, their suppliers, and with the information they need to run their businesses.

Today's hearing will focus on the challenges of providing reliable, affordable broadband access to these rural small businesses. Experience has shown that broadband can bring economic revitalization to small towns by creating clusters of small businesses.

Broadband services can also help farms and farm-related businesses control costs and optimize production. This technology can provide real-time access to weather reports, fertilization guidance, and livestock tracking. Farming communities must maximize their use of high-speed Internet access to encourage further development.

I am concerned that many of our country's rural and agriculture-based communities are not yet fully realizing the potential value of broadband services to their economies. As more of these communities gain high-speed access to the Internet, the next challenge is to help them use this access effectively to help create jobs and sustain growth.

There are many debates going on right now about broadband policy. We must ensure that the needs of rural small businesses are taken into account whenever local, state or federal governments act to change the broadband marketplace. During today's hearing, I hope we will begin a dialogue that will help make sure this happens.

I am very pleased that we have two expert panels here with us this morning, and I look forward to their testimony. I now recognize Ranking Member Fortenberry for his opening statement.

United States Congressman — First District, Nebraska

JEFF FORTENBERRY

May 9, 2007

Statement for Hearing: "Maximizing the Value of Broadband Services to Rural Communities"

Thank you, Chairman Shuler, for scheduling this hearing, and for your remarks. And thanks to our witnesses for appearing today.

This subcommittee is the only committee tasked with the exciting responsibility of encouraging entrepreneurship. Entrepreneurial small businesses are the creators of most of the new jobs in our country, and many Americans are rethinking the old concept of "work" in favor of being their own boss and bringing new products and services to the marketplace. This is a decided trend among young people. And I do believe that the work of this subcommittee can help to remove some of the barriers to creating a more entrepreneurial society.

We are here to discuss broadband Internet access and its importance to rural America. Broadband provides an important gateway to innovation and tools for adapting to the ever-changing market.

A study last year by the Massachusetts Institute of Technology documented that communities where broadband was available enjoyed more rapid growth in employment. Across rural America, businesses in health care, retail, and the agriculture sectors are realizing innovation through growth of advanced broadband services. America, however, has been slower than other nations to adopt this technology. Currently, 36 percent of households have broadband access, but the percentage of broadband usage is lower in rural parts of the country. According to a Government Accountability Office report, the adoption rate of broadband services in rural areas is only 17%, much lower than the national average.

In my own state of Nebraska, we're fortunate that parts of all 93 counties have some form of broadband Internet access. However, over 394,000 Nebraskans live outside population centers, making it more likely that they cannot access this vital service.

In today's hearing we will hear about the potential benefits of having more competitive services for broadband in rural America, and review some of the barriers that stand in the way of such services. In addition, we will review state efforts to pave the way to an expansion of access, and how one state in particular, Kentucky, has become a national leader on this issue. Their effort demonstrates that there is currently no strong definition of what is an un-served area, and illustrates the importance of creating a methodology for defining what areas of the nation are un-served and underserved. Improving the climate for entrepreneurs will depend, in part, on a more nuanced effort by the interested federal agencies to answer these questions and find out what areas of the country need the most attention.

Again, I thank the witnesses, and I await their testimony.

STATEMENT OF
JONATHAN S. ADELSTEIN
COMMISSIONER, FEDERAL COMMUNICATIONS COMMISSION

BEFORE THE
SUBCOMMITTEE ON RURAL AND URBAN ENTREPRENEURSHIP
SMALL BUSINESS COMMITTEE
UNITED STATES HOUSE OF REPRESENTATIVES

MAY 9, 2007

Mr. Chairman, Congressman Fortenberry, and members of the Subcommittee, thank you for inviting me to testify about one of the seminal challenges confronting our Commission and the country: ensuring the ubiquitous deployment of affordable, high speed broadband infrastructure to every corner of this country.

We need to maximize the potential of every citizen to contribute to our social, cultural and economic life through communications, whether they live in major cities or in rural, insular or other high-cost areas, whether they are Native Americans living on tribal lands or residents of economically challenged sections of our inner cities, whether they live with disabilities, whether or not they speak English, and regardless of their income level. I would like to talk to you today about why I believe this is such an important guiding principle for communications policy and a few of the ways we at the FCC and you in Congress can and must work to achieve this ambitious goal. We need to make broadband the dial-tone of the 21st Century.

I am particularly honored to be here because -- as a fourth generation South Dakotan and the first FCC Commissioner from my state, and even from the entire upper Midwest -- I am naturally interested in the important role of broadband as a tool for promoting economic development in Rural America. Early in the last century, my grandfather became an engineer and founded a company that built roads throughout our state. The motto of our family company was "Builders of Better Bridges and Highways." I keep that spirit in mind in my work at the FCC. Just as roads and bridges provide physical links between our communities, our communications networks now bring people together in ways that my grandfather never could have imagined, but that, I'm sure, would have made him smile in wonder.

In some areas of the country, our communications tools have already surpassed the reach of the physical infrastructure to overcome the limits once immovably fixed by distance. I have visited the Bush region in Alaska, above the Artic Circle, where satellite technology, funded through universal service support, connects even some of the most isolated villages to the health and educational facilities of hub cities, even though no roads connect these towns. As we upgrade our nation's communications networks to provide broadband functionality and advanced communications services everywhere, our children will integrate these tools into their lives in ways that we are only beginning to see.

The Role of Broadband for Rural America

By expanding the reach of advanced communications technologies, we can bring new hope to many communities where it is in short or diminishing supply. We are only scratching the surface of the opportunities that these technologies can bring. We stand at the threshold of a revolution in the applications that will ride over this infrastructure. These opportunities hold enormous potential for consumers and workers in small towns and Rural America.

Broadband can connect entrepreneurs to millions of new distant potential customers, facilitate telecommuting and increase productivity. These new connections can create new jobs by allowing businesses to set up remote locations and call centers. Since I have joined the Commission, I have traveled across the country and seen broadband technologies harnessed in rural areas in ways folks back inside the Beltway might never have imagined.

For example, at auction houses across the Midwest, entrepreneurs are using broadband technologies to conduct real time cattle auctions over the Internet. Ranchers from across the country can log in, watch real time video of the livestock and make purchases without leaving their ranches. By putting their livestock up for bid in cyberspace, these auction houses have been able to bridge remote locations, expand their potential markets, and cut the costs of reaching their customers.

Broadband can also unlock transformational opportunities through distance learning and specialty classes that might otherwise be confined within the physical walls of a traditional school. Similarly, telemedicine applications are giving Rural Americans access to diagnostic services, like mobile mammography and emergency services that had been unavailable because of distance, cost, weather, or geography. As we have seen through events like the devastation of Hurricane Katrina, our communications services become even more critical in times of disaster or national emergency, whether as a means of conveying critical information to the public, enabling citizens to communicate with their loved ones, or providing an essential tool for our first responders. Broadband networks are essential to any plan to make emergency networks robust and redundant enough to survive and function in the face of such disasters in the future.

Broadband technologies have the potential to improve the quality of life in even some of most remote and economically challenged communities. On Native American lands, I have seen tribally-owned providers using broadband infrastructure to bring jobs to their communities that serve not only as important sources of employment, but also as training grounds for the young people of the tribes. In almost every small community I visit, I hear how hard it is to develop a workforce with sufficient training in technology. Yet without such workers, it is hard for a small town to develop and oversee cutting edge communications systems. We want people to be able to stay, work, and thrive in the communities where they grew up, yet I often hear that it is harder to keep young people in rural areas these days because they feel a palpable lack of local opportunities. Broadband communications can benefit Rural America in many ways, perhaps most of all by restoring the sense of opportunity that first made Americans venture forth and settle the more remote areas of this country.

As consumers are increasingly empowered to use broadband in newer, more creative ways, the stage on which we all must compete is also evolving into a global one. New telecommunications networks are a key driver of this new global landscape. They let people do jobs from anywhere in the world -- whether an office in downtown Manhattan, a home on the Cheyenne River Indian Reservation, or a call center in Bangalore, India. This trend should be a wake-up call for Americans to demand the highest quality communications systems across our nation, so that we can harness the full potential, productivity and efficiency of our own country. We must give all our towns the tools they need to compete in this new marketplace. If we fail in this, be assured, our competitors around the world will take full advantage of our failure.

Rural America and Global Competitiveness

Keeping our communities connected and ensuring that the latest technologies reach all Americans, including those in remote and underserved areas, are principles that are enshrined in the Communications Act. Meeting these goals will be more important than ever as we enter a new age of global competitiveness.

We've made progress, and there are many positive lessons to draw on, but I am increasingly concerned that we have failed to keep pace with our global competitors over the past few years. Each year, we slip further down the regular rankings of broadband penetration. For small businesses, those in rural areas, and low income consumers, the problem can be even more acute. According to one recent report, seventy-six percent of small businesses in rural areas report no access to terrestrial broadband services. Even more troubling, there is growing evidence that citizens of other countries are getting a much greater broadband value in the form of more available megabits for less money. According to the ITU, the digital opportunity afforded to U.S. citizens is not even near the top: in fact, it is 21st in the world! This is more than a public relations problem. It is a major productivity problem, and our citizens deserve better. Indeed, if we do not do better for everyone in America on this score, then we will all suffer economic injury as a result. In this broadband world, more than ever, we are truly *all* in this together.

Some have argued that the reason we have fallen so far in the international broadband rankings is that we are a more rural country than many of those ahead of us. If that is the case, and since geography is destiny and we cannot change ours, rather than merely curse the difficulty of addressing rural communications challenges, we should redouble our efforts and get down to the business of addressing and overcoming them.

I am concerned that the lack of a comprehensive broadband communications deployment plan is one of the reasons that the U.S. is increasingly falling further behind our global competitors. Virtually every other developed country has implemented a national broadband strategy. This must become a greater national priority for America than it is now. We need a strategy to prevent outsourcing of jobs overseas by promoting the ability of U.S. companies to "in-source" within our own borders. Rural America and underserved urban areas have surplus labor forces waiting to be tapped. No one will work, or work more efficiently, than Americans but many are currently without opportunities simply because the current communications infrastructure is inadequate to connect them with a good job. That situation must improve.

A National Broadband Strategy for All Americans

We must engage in a concerted and coordinated effort to restore our place as the world leader in telecommunications by making available to all our citizens affordable, true broadband, capable of carrying voice, data and video signals. An issue of this importance to our future warrants a comprehensive national broadband strategy that targets the needs of all Americans, including those in Rural America. A true broadband strategy should incorporate benchmarks, deployment timetables, and measurable thresholds to gauge our progress.

We need to set ambitious goals and shoot for real high-bandwidth broadband deployment. We should start by updating our current anemic definition of high-speed of just 200 kbps in one direction to something more akin to what consumers receive in countries with which we compete, speeds that are magnitudes higher than our current definitions.

We must take a hard look at our successes and failures. We need much more reliable, more specific data than the FCC currently compiles so that we can better ascertain our current problems and develop responsive solutions. Giving consumers reliable information by requiring public reporting of actual broadband speeds by providers would spur better service and enable the free market to function more effectively.

We must redouble our efforts to encourage broadband development by increasing incentives for investment, because we will rely on the private sector as the primary driver of growth. These efforts must take place across technologies, so that we not only build on the traditional telephone and cable platforms, but also create opportunities for deployment of fiber-to-the-home, fixed and mobile wireless, broadband over power line, and satellite technologies. We must work to promote meaningful competition, as competition is the most effective driver of innovation, as well as lower prices. Only rational competition policies can ensure that the U.S. broadband market does not devolve into a stagnant duopoly, which is a serious concern given that cable and DSL providers now control approximately 95 percent of the residential broadband market.

There also is more Congress can do, outside of the purview of the FCC, such as providing adequate funding for Rural Utilities Service broadband loans and grants; ensuring RUS properly targets those funds; providing tax incentives for companies that invest in broadband to underserved areas; devising better depreciation rules for capital investments in targeted telecommunications services; investing in basic science research and development to spur further innovation in telecommunications technology; and improving math and science education so that we have the human resources to fuel continued growth, innovation and usage of advanced telecommunications services.

Two other critical steps toward a national strategy, elaborated upon below, are properly channeling universal service and promoting spectrum-based services for Rural America.

Universal Service: Evolving for the Broadband Age

Congress and the Commission recognized early on that the economic, social, and public health benefits of the telecommunications network increase exponentially for all subscribers with the addition of each new subscriber. Federal universal service continues to play a vital role in meeting our commitment to connectivity, helping to maintain high levels of telephone penetration and increasing access for our nation's schools and libraries. With almost a decade behind us since the 1996 Act, the FCC is re-examining almost every aspect of our federal universal service policies, from the way that we conduct contributions and distributions, to our administration and oversight of the fund. As this review has gone forward, I have worked hard to preserve and advance the universal service programs as Congress intended.

Yet, broadband take rates remain comparatively low in Rural America. Among households connected to the Internet, roughly 40 percent of urban households make use of broadband access. In contrast, only 25 percent of rural households with connections to the Internet use broadband. This disparity – confirmed as well by GAO in a 2006 report on broadband -- is perhaps not surprising given that rural residents tend to be less likely to use the Internet, regardless of the technology. The Pew Foundation has suggested that this may be due in part to presence of older and less wealthy populations in Rural America. If this is true, we may expect the same factors to affect the roll-out of broadband in rural areas. But, even if slow broadband uptake truly is only a symptom of them today, we must not allow the lack of broadband access to become an affirmative *cause* of the "graying" and relative impoverishment of Rural America in the future.

Ensuring the vitality of universal service will be particularly important as technology continues to evolve. Increasingly, voice, video, and data will flow to homes and businesses over broadband platforms. In this new world, as voice becomes just one application over broadband networks, we've got to have ubiquitous broadband pipes to carry the most valuable IP services everywhere. Without such broadband networks, IP services can't reach their full audience or capability. The economic, public health, and social externalities associated with access to broadband networks will be far more important than the significant effects associated with the plain-old-telephone-service network, because broadband services will touch so many different aspects of our lives. So, it is important that the Commission conduct its stewardship of universal service with the highest of standards and that we ensure that universal service evolves to promote advanced services, which is a priority that Congress has made explicitly clear.

Wireless: A Critical Source of Broadband Services

One of the best opportunities for promoting broadband, particularly in rural areas, and providing competition across the country, is in maximizing the potential of spectrum-based services. The Commission must do more to stay on top of the latest developments in spectrum technology and policy, working with both licensed and unlicensed spectrum. Spectrum is the lifeblood for much of this new communications landscape. The past several years have seen an explosion of new opportunities for consumers, like Wi-Fi, satellite-based technologies, and more advanced mobile services. We now have to be more creative with what I have described as "spectrum facilitation." That means looking at all types of approaches – technical, economic or regulatory -- to get spectrum into the hands of operators ready to serve consumers at the most local levels possible.

Of course, licensed spectrum has and will continue to be the backbone for much of our wireless communications network. We are already seeing broadband provided over satellite, new wireless broadband systems in the 2.5 GHz band and the increasing deployment of higher speed mobile wireless connections from existing cellular and PCS providers.

During our review of the bandplan in advance of the auction last year of 90 MHz of new spectrum for the Advanced Wireless Service, I pressed for the inclusion of smaller blocks of licenses. I thought that smaller license blocks would improve access to spectrum by those providers who want to offer service to smaller areas, while also providing a better opportunity for larger carriers to more strategically expand their spectrum footprints. Our decision to adopt smaller license blocks was well received by a number of carriers and manufacturers.

The Commission has a historic opportunity in the upcoming 700 MHz auction to facilitate the emergence of a "third" broadband platform that will ensure consumers everywhere the benefits of a high-quality wireless broadband network. As we finalize our rules for the 700 MHz auction, the biggest and most important auction we will see for many years to come, it is critical we build on the lessons learned from our previous auctions to provide a diverse group of licenses so that all bidders have an opportunity to obtain licenses that best match their business plan. While I have supported rules to facilitate the secondary market for spectrum rights and licenses, I think we are best served by providing a wide variety of license sizes at the initial auction when appropriate.

I have also worked closely with the Wireless Internet Service Provider (WISP) community, which has been particularly focused on providing wireless broadband connectivity in rural and underserved areas. I even had the opportunity to host an extraordinary FCC event in my home town of Rapid City a few years ago to highlight the potential of rural WISPs. Unlicensed broadband services can be a big part of the rural solution. Unlicensed spectrum is free and, in most rural areas, lightly used. It can be accessed immediately, and the equipment is relatively cheap because it is so widely available.

We can do even more for rural WISPs and other unlicensed users. I have heard from operators who want access to additional spectrum and at higher power levels. And the Commission has been doing just that. We have opened up 255 megahertz of spectrum in the 5 GHz band -- more spectrum for the latest Wi-Fi technologies – and are looking at ways to increase unlicensed power levels in rural areas.

I also have pushed for flexible licensing approaches that make it easier for community-based providers to get access to wireless broadband opportunities. We adopted rules to make spectrum in the 3650 MHz band available for wireless broadband services. To promote interest in the band, we adopted an innovative, hybrid approach for spectrum access. It makes the spectrum available on a licensed, but non-exclusive, basis. I have spoken with representatives of the Community Wireless Network movement, and they are thrilled with this decision and the positive impact it will have on their efforts to deploy broadband networks in underserved communities around the country.

We have also made spectrum available in the 70/80/90 GHz band for enterprise use. While you may not be familiar with this spectrum block, it can be used to connect buildings with gigabit-speed wireless point-to-point links for a mile or more. Instead of digging up streets to bring fiber to buildings, licensees can set up a wireless link for a fraction of the cost -- and the spectrum is available to anyone holding a license. While others supported an auction, I successfully argued against them in this unique case, because I was concerned that auctions would raise the price of access and shut out smaller licensees. In fact, one company now is installing five links for the city of Sioux Falls. The links will be used for a number of City services, including public works, police and fire departments, as an alternative to fiber.

We are now even looking to allow unlicensed operations in unused television spectrum bands – the so-called "white spaces." It is a challenging proposal, but one that could allow for unlicensed use of spectrum that has exceptional propagation qualities, which can be particularly useful in rural areas. We have an obligation to look at the interference implications of such a proposal, and it will be a major proceeding at the Commission this upcoming year.

Conclusion

Congress has charged the Commission with ensuring that the American public stay well-connected and well-protected, directing us in the very first section of the Communications Act with making available to "all the people of the United States" rapid, efficient Nation-wide communications services. That starts with a continuing commitment to connectivity, and nowhere is this more important than in Rural America.

If the horse and wagon were the key tools that allowed my ancestors to settle the west, broadband networks will be a big part of maintaining and restoring the vitality of our rural communities in the future. Let us face this new frontier of silicon and fiber as bravely and resourcefully as they did the original frontier of forbidding forests and vast prairies. If we do, I know we will experience similar success and the proper place in the history of American progress. For the sake of ourselves, our children, and this great country, may we be as bold and successful in our own pioneering endeavor as they were in theirs. Thank you for your leadership on rural broadband and for the opportunity to testify before you today.

Testimony of BalsamWest *Fiber*NET, LLC

to

Subcommittee on Rural and Urban Entrepreneurship

of the Committee on Small Business of the United States House of Representatives

"Maximizing the Value of Broadband Services to Rural Communities"

May 9, 2007

Mr. Chairman and distinguished Members of the Subcommittee,

Thank you for allowing BalsamWest FiberNET to present testimony to this subcommittee concerning the importance of broadband services to rural areas and the small businesses and entrepreneurs who live and work there.

BalsamWest is a certificated Interexchange Carrier and Competitive Local Exchange Carrier in North Carolina, Tennessee, and Georgia. BalsamWest is a unique company and an example of the success that can be achieved through collaboration, determination, and innovation.

BalsamWest was formed in 2003 to serve the critical need for access to reliable, affordable high-performance broadband infrastructure and services in the southwestern region of the Southern Appalachian Mountains.

BalsamWest's testimony concerns the unusually severe challenges faced by a rural mountainous region in obtaining reliable, affordable, high-performance broadband infrastructure and services, the difficulties of telecommunications and cable companies in serving the needs of the people in the region, and how—of necessity—people from within the Southern Appalachian Mountain region worked collaboratively to overcome the problem on their own—with stunning success.

The area served by BalsamWest is one of profound beauty with terrain ranging from the highest mountains east of the Rockies to deep, shaded gorges where trout-filled streams turn into rushing whitewater rivers. One of the most geographically isolated and economically challenged areas in southern Appalachia; this area is surrounded on all sides by the 5,000-foot peaks of the Blue Ridge and Smoky Mountains. Small communities and towns have grown up wherever level ground was formed by rivers carving valleys through the mountain ranges. Communities are widely separated by steep, winding roads to the next river valley. Population in the region has historically been sparse, but is growing due to an influx of retirees, second-home owners, and resort developments attracted to the scenic beauty and lifestyle of the region. Tourism is a major economic force in the region.

The culture of the mountain people is rich, yet encapsulated. In the history of the region, there have been many challenges to overcome in obtaining basic infrastructure. In the more recent past, there have been barriers to economic prosperity in the face of agricultural decline, off-shore job flight, and the seasonal nature of tourism. Far from major trading centers and isolated by formidable terrain, mountain people have had to independently overcome many challenges through collaboration, determination, and innovation.

Coupled with these challenges has been the simultaneous emergence of a new economy powered by 21st Century technologies and knowledge resources. The new economic order presents a wealth of opportunity for those who have access to these resources— increased economic opportunities, higher living standards, better schools and health care, stronger communities, and more meaningful participation in government and public life—and a widening gap for those who do not. Economic development in our region increasingly depends less upon landing a single, blue-chip industry and more upon how well we have prepared our labor force and supported out small businesses and entrepreneurs to enter and thrive in a knowledge-based economy.

To create a new economic reality for itself, our region must have:

A Poised Infrastructure ➔

1) Access to reliable, affordable high-performance broadband telecommunications infrastructure and advanced services at competitive prices;

➘

2) Connectivity to major metropolitan areas outside the southern Appalachian mountains, overcoming long distances to major trading centers and jobs with high-speed connectivity;

A Poised Workforce ➔ 3) Community-based digital literacy campaigns as well as seamless, comprehensive training and re-training opportunities;

A Poised Market ➚ 4) A process for identifying, engaging, readying, and supporting the local workforce, small businesses and entrepreneurs to transition to—as well as perform and compete in—a global marketplace; and

➘ 5) Innovative and collaborative approaches to business market expansion to a more diversified and robust national and global platform.

By 2002, the lack of affordable access to reliable, high-performance broadband connectivity in the region had become a serious impediment to effective provision of health care services, education, government services, and to economic development. The needs of existing businesses for connectivity to modernize and expand were great and economic development recruitment activities were hampered when prospects inquired about access and cost of broadband connectivity. Prospective buyers of second homes in resort communities wanted access to "Triple Play" and "Quadruple Play" services available in the largest cities in the U.S.: IPTV (full high-definition video entertainment over Internet protocol delivered to the home over fiber-optic cable), VoIP (unlimited voice calling over Internet Protocol), digital surveillance, and high-speed Internet service of at least 8 megabits per second.

Cell phone coverage in the mountains is also limited by the terrain. Tower siting is difficult due to the mountains, and the high percentage of publicly-owned protected natural areas where towers would spoil the scenic beauty. While mountain people had grown used to being "out of range" in many places in the region, people moving to and visiting the region were not used to being out of touch for such long distances. While people in cities are rapidly replacing their landlines with cell phones, many people in the mountains do not have that option.

While it would seem logical to place the blame on local telephone and cable companies for failing to serve the needs of the area, it would be unfair to do so. The local management and employees of these companies live in the area, their children attend schools here, and they must obtain health care and government services from local institutions. They have the same needs and hopes as all the other residents of the mountain communities, and they do their very best with the infrastructure resources available to them. The problems are at a higher level:

- The law of supply and demand fails in areas that are so extremely difficult and expensive to serve due to formidable terrain barriers and low population density.

- From a policy standpoint, the policies and regulations that Congress and the FCC pursue do affect the ability of competitive companies to deploy and provide service. In the past few years, the FCC has not pursued policies that promote competition and protect consumers, with especially adverse results in rural isolated areas that are difficult for any carrier to serve. Instead, the FCC has favored policies that specifically benefit the large incumbents. These large, publicly-traded companies, with shareholders expecting returns comparable to investments in other industries, cannot justify the necessary investments in advanced infrastructure and services that are so critically needed in areas like the southern Appalachian Mountains. Many areas of the rural region have had no other choice but to be served by the large incumbent carrier, which in turn cannot justify investment in the area. The result of these policies is that our businesses—which are comprised mostly of small businesses and entrepreneurial ventures—have been critically affected. We have fallen far behind in the new, global economy, and our school children and rural businesses are unprepared and unable to compete.

- Federal and state incentive, grant, and loan programs have restrictions that either prevent investments in the infrastructure, or are too complicated, cumbersome, and slow to be of benefit.

The fiber-optic infrastructure serving traffic in and out of the region was deployed when population in the area was still growing slowly. Growth in population; the now critical importance of Internet access to residents, businesses, government, education, and health care; imaging applications for education and health care, and the rising demand for high-definition video and Triple and Quadruple Play services by homeowners had driven demand to the point at which existing infrastructure was becoming too constrained to serve the demand for higher bandwidth services.

More importantly, reliability of service had become a major issue by 2002. The region is served by a number of telecommunications companies that interconnect their fiber-optic networks to transport traffic to remote switching facilities. In some areas, these interconnections form a lateral line, not a ring. This is an important concept, because a single cut of an underground lateral fiber-optic line can cause widespread outages. In a fiber-optic ring configuration with high-end electronics, traffic is automatically rerouted in the other direction on the ring when an outage is detected.

In 2002, there were 8 outages in the western North Carolina area of the region. One of these outages took down communications over landlines, cell phones, and Internet for several hours over four counties.

Clearly this created problems for residents and businesses in the region. Millions of dollars of electronics funds transactions of two local businesses in the area, Drake Software of Franklin, North Carolina, and Harrah's Cherokee Casino in the Cherokee Indian Reservation in Cherokee, North Carolina were affected.

Why couldn't the telecommunications companies serving rural communities in this area expand the capacity of the fiber-optic infrastructure and create a ring to solve the reliability issue?

It was too expensive to do so in light of the terrain barriers – extensive rock and steep terrain – and population density was too low to provide a return on investment that is expected by telecommunication carriers. A large share of the cost of expansion and building redundancy would have to be passed on to the people in the region.

In 1998, the people and organizations in the Western North Carolina area of the region began meeting with Southwestern Community College of Sylva, North Carolina to share concerns and search for answers to the dilemma.

Southwestern Community College

Southwestern Community College, a longtime community and technology leader in the western North Carolina area of the southern Appalachian region, had offered distance learning in the early 1990s to the region's schools in its service area. Needless to say, distance learning curricula are critically important to rural schools that cannot afford to hire a teacher of advanced math and science programs for only a few children. Young people graduating from rural high schools and entering college often found themselves at a competitive disadvantage with students from metropolitan area high schools with large enrollments, where college level math and science programs are a common offering.

Southwestern Community College created an interactive instructional television network, Community Link. This closed-loop, analog-based system was designed to deliver full-motion video and audio among 13 sites—including the community college sites, four public school districts, and a regional university. This network was especially beneficial to public high schools previously unable to offer low-enrollment, high-cost, advanced course work that would not otherwise be available to small rural high schools.

As capacity on the serving carrier's network became constrained, and services became more expensive to provide, Southwestern was faced with changing its analog, full-motion video system to compressed video to save bandwidth cost. The quality of video using compression technologies at that time was not high-quality, and the learning experience for students and teachers suffered. Southwestern's experiences with high cost and service quality meshed with others' in the community. Southwestern realized that the isolation of the region and the high cost of connectivity out of region would continue to exclude it from the resources needed to compete in the new economy. They understood that the region must gain the ability to reach outside the confines of the mountains—perhaps its only hope for joining mainstream America in the fruits of the 21st Century. Finally, the College was aware of other applications, rich electronic content, and on-line services that would add tremendous value to the lives of the people living and working in the region. The College and the region would not be able to seize these opportunities without better telecommunications infrastructure.

In 1999, Dr. Cecil Groves, President of Southwestern Community College and current Community Interest Director of the BalsamWest Managing Board, initiated the Appalachian Access project to examine the needs of the region, to identify the exact nature of the issues affecting rural infrastructure deployment, and develop and assess strategies for resolving the issues. The design of the project comprised an examination of the business issues, regulatory, technological, and political factors affecting rural broadband infrastructure deployment. Mr. Jim Campbell, Vice President for Information Technology, and Mrs. Laura Pennington, Community Resource and Development Director, were responsible to inform and manage this initiative on behalf of the institution. The College partnered with Sherry McCuller, Managing Director of The Institute at Biltmore, a regional nonprofit research and planning organization and current Manager of BalsamWest FiberNET, to conduct research in the four project focus areas, develop and assess strategies in light of the factors affecting infrastructure deployment in a mountainous rural area.

The project was a groundbreaking initiative to develop in-depth knowledge and understanding of the highly regulated and complex world of telecommunications infrastructure and service delivery, and a grassroots effort to lower the cost and increase availability of broadband telecommunications services in rural western North Carolina.

The Appalachian Access partners discovered that in remote, rural areas with terrain barriers, such as communities in mountains or islands and peninsulas, the highest cost component of high-performance broadband service is more often not in the "Last Mile"—the portion of the community network between the serving carrier facilities and customer premises—as is frequently the case in other rural communities. Instead, the high-cost component is often in the "Middle Mile"—the regional network between the local carrier facility and out of area "traffic aggregation hubs". In the isolated, low population area, traffic was transported—sometimes long distances—to a remote location, where it was combined with traffic coming from other areas in the region and then transported to interconnection points where the traffic hopped onto major US long-haul networks that crisscross the country.

Internet traffic is transported to Internet Points of Presence or "PoPs" over these long-haul networks. PoPs are facilities where fiber-optic network owners physically interconnect their fiber to exchange traffic and drop and pick up services.

The closest Tier One (largest) PoP to the southwestern North Carolina and northern Georgia area of the southern Appalachian region is in Atlanta, Georgia. Two 40-story buildings on Marietta Street provide physical interconnection facilities for the fiber optic networks of almost every major carrier in the US and some international carriers to interconnect and exchange traffic and drop off or pick up "content" such as Internet Service, Internet Protocol Television ("IPTV") and Voice over Internet Protocol ("VoIP"). This content is transported over the fiber-optic networks to other carriers and on to consumers.

The importance of a Tier One PoP cannot be overemphasized because content can be purchased at these PoPs at the lowest possible cost. A Tier One PoP spurs rapid economic growth in a surrounding area as businesses and other types of organizations and institutions locate nearby to reap the benefits of proximity to lowest-cost transport and content.

Transport on the middle mile network in the rural Southern Appalachian region was priced by the mile, and that is the primary reason why the cost to the area was up to ten times higher than higher-quality services offered in metropolitan areas.

While Last Mile services were also high-cost, the cost and service quality of the Middle Mile network represented a much greater issue. The middle mile network would have to be expanded and improved, or a new network would have to be built. This was a daunting prospect in terms of cost and time to completion.

To spur investment in this component of infrastructure, the Appalachian Access partners developed a strategy to aggregate regional demand for broadband services, and combine the demand with incentives in the form of grants or public subsidy. The demand aggregation phase took almost a year and involved 129 local businesses and institutions who all agreed to aggregate their demand in a volume purchase of services.

Project principals began the negotiation process with 21 prospects in 6 categories:

- Transport providers (fiber providers)
- Applications & managed network providers
- Cable providers
- Supply brokers
- CLECs
- ILECs

All of the vendors were provided with the following list of criteria that would be used by project principals in selecting the vendor:

- Reduced prices for advanced telecommunication services for end users - at parity with adjoining urban areas;
- Enforceable service level agreement;
- Blended rate pricing (one price for everyone on network), rolling rate based on volume;
- Flat rate, not distance sensitive, middle mile pricing.
- Reduced local loop charges (last mile);
- Length of contract term (longer-term contracts allowable for high-bandwidth fiber customers);
- Redundancy to at least two national long-haul interconnection sites;
- Number and type of peering relationships;
- Degree of fiber infrastructure in proposal;
- Time required to offer services;
- Time required to provide infrastructure; and
- Other value-added offerings.

Project principals structured the following vendor incentives:

- Marketing and aggregation of demand for telecommunications services through a regional nonprofit "Master Demand Aggregator", the Western North Carolina Knowledge Coalition. These services would be provided free of charge to the winning vendor.
- Significant network construction subsidy.
- Continued resource development by the Appalachian Access team to raise additional sources of funding for the required network construction above the subsidy.
- Low-cost expansion and strategic positioning in a growth area. (Asheville, North Carolina is within a 2-hour drive of 7 of the top 200 growth cities in the US as listed by Forbes Magazine in 2001).

- National press and a replicable rural modal for the vendor to expand its customer base throughout rural America.

The US economy began a precipitous decline as early as February of 2001—and with spectacular ill timing the telecommunications sector led the plunge. With venture capital drying up for telecommunications companies and stock prices dropping at alarming rates—project principals had to rethink its list of selected vendors—in light of several factors:

- Financial stability (cash levels replaced revenue levels and market share as our focus);
- Potential to continue to attract venture capital for approximately 18 months on declining share prices; and
- Willingness to invest in capital equipment to expand into rural markets as liquidity and risk management became key factors for economic survival.

In the most fundamental terms, telecommunications companies began to focus on cash flow and improvement of liquidity ratios. Strategic moves requiring capital expenditures moved to the bottom of the priority list for all but the most cash-rich companies. As one venture capital company analyst remarked, "If a venture-backed CLEC has to spend one dime on capital equipment to expand its market territory, the deal isn't going to happen." The future of the project seemed dubious.

In the end, there was one viable vendor candidate who vied keenly for the region's business because it represented an attractive strategic positioning opportunity, as well as a way to deploy a lower cost build through the area to connect from its operations in Florida to a large new business customer in New York.

In the negotiations with this vendor, project principals worked tirelessly through every aspect of the business model—which was one of a for-profit entity looking for an internal rate of return, seeking to contain risk, minimize upfront investment, and maximize upside potential, while maintaining asset liquidity in a dangerous marketplace where Chapter 11 and divestiture loomed large.

The vendor agreed to invest significant capital in infrastructure deployment in the area in return for incentives, and assigned senior and technical staff to work side-by-side with the Appalachian

Access team to conduct the design and engineering of the fiber-optic middle mile network ring and electronics, to work through the interconnections at local carrier facilities, to develop the cost of a dedicated connection to a Tier One PoP to reduce the cost of content, to develop the cost of the demand for services identified by Appalachian Access, and to develop the structure of the transaction.

This work spanned almost a year

Tragically, the devastating terrorist attacks of September 11, 2001 occurred as the transaction was nearing final completion, and the vendor—like most companies in the U.S.—became concerned with cash conservation, liquidity, and risk. Ultimately, to the great sorrow of all involved, the vendor could not take the financial risk of entering the mountain region marketplace at that time. The Appalachian Access partners remain grateful to this vendor to this day for its tireless commitment to making an innovative plan work for the region and itself, and for the hard work of the vendor management and staff.

Appalachian Access partners turned to the public and nonprofit sectors for funding the deployment of the network and plan. Just as the private-sector viewed investment of capital into the mountain region telecommunications infrastructure as high-risk in terms of return on investment, so too did the public and nonprofit sectors. The project was viewed as too expensive and too risky by these sectors. The same factors that prevented deployment of a redundant, high-performance network by the private sector were also barriers to the public and nonprofit sectors.

After years of work, the Appalachian Access partners had to begin again, but this time they were armed with a comprehensive plan, knowledge, and expertise. The work of the Appalachian Access principals led to the conclusion that the rural mountain region would not be served with comparable infrastructure and services as metropolitan areas by the private-sector unless demand justified the investment, and even aggregated demand, incentives, and strategic advantage were not enough to attract investors in a downturn in the U.S. economy in the aftermath of the terrorist attacks on September 11, 2001. Western North Carolina needed regional private-equity investors willing to develop and own a middle mile ring that would traverse the entire region, and for economic reasons other than return on investment in and operation of a telecommunications company or cable company.

In 2003, Ms. McCuller resigned as Managing Director of the Institute at Biltmore to concentrate her efforts full-time on developing a regional private equity funding model for infrastructure in the region. Ms. McCuller and her family, like the other members of the Appalachian Access team and BalsamWest FiberNET, live in southwestern North Carolina, and she understands first-hand the critical importance of the high-performance infrastructure to the region. She believed that her experience gained through a 20-year career with a regional investment bank in Charlotte, NC, where she served as CIO and CFO before retiring in 1992 would be important to the continuing effort and she was deeply committed to its success. Ms. McCuller formed Peregrine Management Partners to partner with Southwestern Community College and continue the work.

Regional Collaboration

By 2003, the partners knew that the only real hope for the region was to raise private equity from investors within the region whose businesses and operations depended heavily for success upon high-performance broadband and reliable infrastructure. Two organizations who were seriously concerned with the impact of the situation on their own operations responded to Southwestern's call for regional support, and BalsamWest FiberNET was born.

Drake Software

Drake Software of Franklin, North Carolina met with Southwestern Community College to explore the feasibility of developing such a network. Drake Enterprises is a family of 13 companies in the region, whose flagship company, Drake Software, was created by Phil Drake – a highly successful entrepreneur with deep roots in the region. Mr. Drake's company has grown from his development of an innovative software program to transmit tax return information electronically from accounting firms to the I.R.S., to the second largest electronic tax filing company in the U.S and a clearinghouse for approximately $5 billion of electronic funds transactions associated with income tax returns. Dnet Internet Services, the Internet arm of Drake Enterprises, was hard pressed to maintain reliable connectivity to major Internet PoPs during tax season in light of the network outages in 2002, and its cost of service was much higher than the cost of comparable service in metropolitan areas. Mr. Drake did not want to move his business – which has created over 530 jobs in the area – out of the area in order to obtain affordable access to needed high-performance broadband services.

Mr. Drake and his management team are committed to improving conditions in the region for their business, their families, and all of the people in the region whose prosperity is so intertwined. Thanks to the work of the Appalachian Access initiative, Mr. Drake was able to clearly see the problem, the necessary action, the cost, and the potential return to his business from cost savings and reliability improvements. He knew that from a purely business prospective, he was faced with moving his business out of the region or solving the problem independently. He was well aware of the benefits to other small and medium-sized businesses in the region, as well as the benefits to education, health care, and government services in the region. While the return on investment on development and operation of a telecommunications interexchange carrier middle mile network and connectivity to a major metro Tier 1 PoP were, on their own, not compelling—the benefits to his core business, his subsidiaries and all of the businesses and organizations in the region were clear.

In September 2003, Mr. Drake committed to an investment to begin the deployment and operation of the BalsamWest FiberNET network, and appointed two members of his senior management team to represent Drake Software in BalsamWest FiberNET: Mr. David Hubbs, Director of Dnet Internet Services, which serves Drake Enterprises and also operates as a local ISP, and Mr. Tim Hubbs, President of Drake Enterprises. Both David and Tim Hubbs and their families have deep roots in southwestern North Carolina. David Hubbs served BalsamWest as its first Chairman of the Managing Board.

Construction of what is now a 300-mile network through the southern Appalachian mountains connected to Atlanta, Georgia, commenced with construction of a the first phase of the ring connecting Franklin to Webster to Sylva, North Carolina and a spur to Cullowhee, North Carolina to allow the regional university, Western Carolina University, to also benefit.

Eastern Band of Cherokee Indians

Southwestern Community College's distance learning network serves the Qualla Boundary of the Eastern Band of Cherokee Indians. Tribal Government, headquartered in Cherokee, NC, and Tribal businesses, merchants, hospitals, schools, and residents had also been deeply affected by high-cost services, limited capacity, and reliability issues. Tribal Government began examining the Appalachian Access assessment and plan, and began discussions with Drake Software and Dnet Internet Services to explore the BalsamWest partnership to pool capital to deploy and operate an advanced middle mile network.

Principal Chief Michell Hicks and the Tribal Council authorized Mr. Brandon Stephens, Tribal Planner and Grant writer, and current Chairman of BalsamWest's Managing Board to represent the Tribe's interests in the BalsamWest venture.

Economic and Social Conditions:

Despite the scenic beauty of the Great Smoky Mountains National Park and the Nantahala National Forest, western North Carolina is characterized by poverty and diverse social needs. The multi-generational residents, largely Cherokee and Scotch-Irish, have a similar experience of independence and self-reliance that is characteristic of most mountain people. They have much in common with other Appalachian people faced with stagnant economies. Because the town of Cherokee serves as the eastern gateway to the Park, much of the area has a long history of serving traditional tourists. Yet, the area has limited employment opportunities for those with special skills, so a serious out-migration of young, educated residents has been all too typical. Much of the current employment is seasonal and at low hourly wages. Unemployment rates increase in the winter months to sometimes as great as six times the national average. These economic conditions have contributed to the need for social services that are seriously underfunded in these poor counties.

Separate statistical information related to the economic and social conditions for each sector of the population within the Qualla Boundary and Eastern Band of Cherokee Trust Property follow.

Census blocks are varied between 1,407 and 4,696 in population and the percentage in poverty ranged from 24 to 38 percent;

Beyond the large number of people living in poverty, it is also important to note the high percentages of workers who are considered underemployed in the area. The centrifuge in saying that an individual is underemployed is that residents are overqualified for the available jobs in the area. The underemployed, those earning $7.00 to $8.00 per hour or less, were identified by North Carolina County:

	Graham	Jackson	Swain
Persons earning $7.00 or less per hour:	676	3,580	2,508
% earning $7.00 or less per hour:	30.1 %	34.2%	48.2%

Tribal members have discovered the difficulty of establishing a business. Those who have the ability to operate a business have discovered that receiving financial assistance through recognized banks, financial institutions, and private financiers is not a guarantee. These organizations are ordinarily unwilling to take the risk of lending money to business owners on the Trust Property. Like most Native American Reservation or Trust Property, non-enrolled members are unable to own property. The financiers are unable to collect on a defaulted loan because they are unable to own the property. Smaller loans for automobiles or home furnishing are made because they are typically able to repossess them. The Eastern Band of Cherokee Indians Tribal government is able to repossess or take eminent domain over Tribal Trust Land.

The Eastern Band of Cherokee Indian trust lands encompass more than 88 square miles. The majority of Trust Property covers parts of Swain and Jackson counties with smaller tracts in Graham and Cherokee counties and approximately 250 acres in Haywood County. There are approximately 9,500 enrolled members of the Eastern Band of Cherokee Indians living on trust lands compared to the near 14,000 total enrolled members. Of the enrolled members living on trust property, 20 percent were judged to be living below the national poverty level. The population is about evenly split between males and females.

Employment is strongly tied to tourism, which is the number one factor affecting the economy of the Cherokee Indian. Summer unemployment levels have dropped as low as five percent but the situation changes dramatically during the winter months when at least 30 percent of the Indian work force is unemployed. In June 1998, 947 or 14.7 percent were unemployed out of a tribal labor force of 6,432. The United States Department of the Interior and Bureau of Indian Affairs has recently calculated the average unemployment rate to be 13.15%.

The situation has been especially difficult in the Swain County and Graham County portions of the Reservation where several industries in the neighboring community of Bryson City have either closed or faced severe employment cutbacks. In Graham County business and industry has been limited because of geographic barriers. The Appalachian Regional Commission lists Graham County as the only distressed county in the state of North Carolina, out of the near 30 counties it serves. To be a distressed county, according to the Appalachian Regional Commission, the sustained unemployment rate must be 8.6% or higher, a 19.7% or higher poverty rate, and a per capita market income of $12,934.00 or less.

Current tribal per capita income was estimated at $10,000 per year. Median family income was judged to be $15,956.00 per year. The State reports the median family income as $26,647.00 and the national average is $16,450.00.

The fight for property and resistance to ethnic cleansing is no longer a fear for the Cherokee. The fear now is loss of identity, tradition, and disease.

In recent years, the Eastern Band of Cherokee Indians have made great efforts to encourage enrolled members to learn the Cherokee language and alphabet. The Tribe is also working in a diligent fashion to restore and recreate design, tradition and cultural significance to many things. Restoration and renovation to places like the Cherokee Ceremonial Grounds, downtown business district, and the manner business is being conducted.

Diabetes eclipses most concerns. The Eastern Band of Cherokee Indians is experiencing a crisis in diabetes prevalence that has reached epidemic proportions. Of the 14,000 enrolled members, 1 out of every 3 persons suffers from this incurable disease. As a result of the high rate of the disease, the number of amputations has tripled in the last ten years.

The last count of Cherokee with diabetes eclipsed 1,600. Ten years ago there were 681, a number that has more than doubled. In the total number of Tribal diabetic population exists a greater number of children who have developed adult-type or type two diabetes. The real tragedy is that just when these children have reached young adulthood, they face the risk of developing diabetes complications such as leg amputations, kidney failure, and blindness. These are complications that a normal adult diabetic would face at senior adult ages. There are six times as many people with diabetes who have early and/or late stage kidney failure, heart disease, or leg amputations than someone who does not have diabetes.

Amenities in the Qualla Boundary portion of trust property include an elementary school with 600 students in kindergarten through eighth grade and a high school with 200 enrolled. There are as many as 176 students enrolled in Cherokee Head Start. There is a satellite campus of Southwestern Community College on the Qualla Boundary. The main campus is in Sylva some 15 miles away. There is also a satellite campus of Western Carolina University whose main campus is 18 miles away in Cullowhee.

Snowbird in Graham County is the most remote part of the Eastern Band of Cherokee Indians trust property. The population density is fewer than 25 people per square mile. The largest town is Robbinsville with about 800 citizens. 70% of the county is federally controlled land, a limiting factor for increasing the tax base. Ninety-six percent of county land is classified as forest, overwhelmingly as timberland with 60 percent counted as national forest. In the most recent Census the county had almost 8,000 people, which are 800 more than the 1990 Census.

In 2004, the Eastern Band of Cherokee Indians committed to an equal partnership with Drake Software to complete the middle mile network and operation of BalsamWest FiberNET. The Eastern Band also committed funding, as did Dnet Internet Services, to the development of community local loops through the communities of Franklin and Cherokee connecting to the BalsamWest regional ring. The Principal Chief of the Eastern Band, Chief Michell Hicks, appointed the current Board members of BalsamWest representing the Eastern Band of Cherokee Indians: Mr. Brandon Stephens of Sylva, NC, current Chairman of BalsamWest's Board, and Mrs. Barbara Vicknair, former Cherokee County Commissioner from Murphy, North Carolina. Mr. Stephens and Mrs. Vicknair are enrolled Members of the Eastern Band of Cherokee Indians, and their families have lived in the region for generations.

The Eastern Band of Cherokee Indians made the decision to invest in the fiber-optic network infrastructure of BalsamWest due to network outages in the region that disrupted communications. The Tribe lost millions of dollars during these outages, and the loss for the region as a whole was estimated at $72 million. To improve service and offer residents on tribal lands high-performance broadband service and Internet access, the Tribe made the decision to partner with Drake Software.

The Tribe has begun to realize the benefits of the regional fiber-optic network and its community fiber-optic loop. Combining the low-cost, high-performance broadband infrastructure and services with tax incentives and a location in a region surrounded by high growth, the opportunities for development became evident. The Tribe has been exploring economic development outside of tourism, including call centers, software development, health care provision, and national entertainment.

To prepare the community for such development, the Tribe has provided high-performance connectivity for the Cherokee Indian Hospital to send and receive images, telemedicine applications, and video-conferencing. The housing department is exploring amenities such as Voice over Internet Protocol telephony ("VoIP"), security and surveillance systems, high speed Internet service, remote control systems for environmental monitoring and control, and Internet Protocol television ("IPTV"). The Tribe is also building on the new connectivity for Tribal schools for distance learning, video-conferencing, ultra-high speed connectivity to other schools and colleges, high-speed Internet services, and a host of education learning tools and content now possible with the installation of fiber-optic connectivity to the Cherokee school sites.

The challenges and barriers for the Cherokee in this new work of technology is that they lack a technically trained labor force due to the years of out-migration as described earlier and the access to financial resources and technical assistance to continue to move the fiber optic project forward. The work will be slow in the beginning to build a new economy and one that will attract the best and brightest, who now leave the Reservation to seek an education and never return. The Tribe is also learning that in the beginning it must rely on its own resources as there are no available state resources from North Carolina. Federal support programs are difficult to navigate and support, and the process is long. Outdated policies prevent the Tribe from taking part in active programs that support the advanced services and infrastructure needed by the Tribe and the region. Generally the programs are designed to meet the needs of regions that live in extreme conditions where the population is low (100-200 city population); the poverty level or unemployment level is 10 times the national level; and where an almost absolute zero level of connectivity exists. Broadband connectivity is currently defined as 200 kilobits per second of bi-directional bandwidth, but this bandwidth will not support high definition imaging and programming, and it would be difficult to conduct interactive, distributed work at this speed. Requirements currently imposed are difficult to meet in most rural areas like the Cherokee Indian Reservation, but the area still needs support in the form of resources to expand and leverage what has been begun by BalsamWest.

Deployment and Operation of BalsamWest FiberNET

In 2004, Ms. McCuller invited John Short to join Peregrine Management Partners to provide needed expertise in telecommunications service provision. In December, 2004, the Managing Board of BalsamWest FiberNET retained Peregrine Management Partners as the company's Manager. Peregrine reports to the BalsamWest FiberNET Managing Board.

Construction moved forward rapidly in 2004. BalsamWest collaborated with a scenic railroad in the area – the Great Smoky Mountain Railroad – to share the railroad Right of Way through some of the most difficult terrain in the entire area – from Dillsboro over Fontana Lake and through the Nantahala Gorge, areas with very steep terrain, sections of almost solid rock, and trestle bridges over Fontana Lake created by the historic Fontana Dam. Great Smoky Mountain Railroad's collaboration allowed BalsamWest to save costs in support in the region, and the Railroad was able to gain access to infrastructure along its railway to enhance its own and its passenger's communications capabilities.

The N. C. Department of Transportation supported the effort to bring this badly needed infrastructure to the region. Mr. Conrad Burrell, the regional representative on the State Commission, communicated the importance of this effort for community development and economic development in the area to the Commission members in the State Capitol.

In 2005, BalsamWest connected its first customer: the West Care Hospital System, which was able lowered its transport costs between remote, rural health care facilities by 25 times, and decreased transport time of critical medical imaging between facilities from 30 minutes (by car) to 12 seconds. The President of the Hospital, Mr. Mark Leonard, and the IT director of the System, Mr. Shawn Remacle, understood the benefits to the hospital and to the region of the network. Westcare's innovation and collaboration allowed BalsamWest to use facilities in two of its hospitals to create colocation facilities for the communities it served. BalsamWest installed state-of-the-art electronics and rack space for others in the community to interconnect. As a consequence of Westcare's collobation, the remote facilities in the WestCare system are located directly on a 300-mile underground network connected to Atlanta, Georgia. The WestCare system has been able to lower its cost of transport service ***by 25 times***, transmit critical medical imaging ***in seconds***, gain access to new and valuable medical applications and Internet service, and benefit the communities and the BalsamWest network at the same time.

Also in 2005, BalsamWest worked with Blue Ridge Mountain Electric Membership Cooperative, a Young Harris, Georgia-based organization serving Clay and Cherokee counties in southwestern North Carolina and Towns and Union County communities in northern Georgia. Blue Ridge Mountain EMC and its Executive Director, Mr. Joe Satterfield, and Community Development Director, Mr. Erik Brinke, have long served the rural region with electric infrastructure and services and are deeply committed to community development in the area. Blue Ridge Mountain EMC wished to use its electric utility poles and facilities to offer Internet services to underserved communities in its electric service areas. BalsamWest tailored its construction design in this area to meet the needs of Blue Ridge Mountain EMC, and Blue Ridge Mountain EMC obtained long-term ownership of strands of fiber being constructed through the region. Blue Ridge Mountain EMC provided facilities in its Young Harris, Georgia headquarters for BalsamWest's use in installing its electronic equipment, and BalsamWest created a colocation site open to the community for interconnection. BalsamWest was able to obtain a facility with excellent electric utility service and back-up electric service, and Blue Ridge Mountain EMC's headquarters facility was then located directly on the regional middle mile ring, with off-net service to Atlanta. Blue Ridge Mountain EMC, BalsamWest FiberNET, and the rural communities all benefitted tremendously from this arrangement.

In 2006, BalsamWest deployed local community loops connecting to the middle mile network in Sylva and Bryson City, North Carolina, and CopperHill, Tennessee. In Sylva, BalsamWest provided a low-cost spare duct to a local Internet Service Provider and CLEC: Metrostat Communications. Since that time Sylva-based Metrostat has deployed copper, fiber, and wireless access to the downtown Sylva area and installed a free Wi-Fi system for the town. BalsamWest's investment in underground fiber-optic infrastructure began to benefit local last mile service providers. In Bryson City, BalsamWest provided duct and fiber to the Swain County Government and Bryson City municipal government for use in connecting government sites to save cost and improve capacity and transmission speed.

Also in 2006, BalsamWest completed the first phase of a project in which it takes great pride. This project, known as "WNC EdNET" is the first of its kind. BalsamWest worked collaboratively with Southern Pipeline, Drake Enterprises, the Eastern Band of the Cherokee Indians, and Blue Ridge Mountain Electric Membership – collectively known as the "ASAP Partners" ("Advancement of Southern Appalachian Prosperity Partners") to knit its middle mile network fiber together with the fiber belonging to the other ASAP Partners and low-cost construction services where new construction was required. The network design connects all 70 K-20 educational institutions in a 6-county rural school district and the Cherokee Indian Reservation. This network is a private, educational network connecting the public and charter K-12 schools together with one another and the two community colleges and regional university serving the region.

BalsamWest has assisted the public and charter schools in southwestern North Carolina in leveraging BalsamWest's investment of $16 million in rural fiber-optic infrastructure. The ASAP Partners, working collaboratively, offered this network to the schools in a competitive bid at ***more than $60 million lower than the second lowest bid***.

The schools are being connected together on their own private fiber-optic network with virtually unlimited capacity, which is controlled by the schools. This network allows the schools to share scarce resources and work together, if they wish, to purchase services in volume. More importantly, it allows the two community colleges and regional university to connect directly to all the schools, some of which are located in remote, isolated areas, to provide real-time high definition distance learning to the classroom. The high capacity network will support concurrent, interactive programming from the community colleges and university to multiple schools. This is critical for rural schools that cannot afford to hire teachers to offer subjects to only a handful of students. Metropolitan area schools with high enrollment routinely offer college level math and science classes. Rural students are often at a competitive disadvantage in science and math upon entry to college.

The schools had initial funding of only approximately $2.8 million upon award of the bid, awarded through the Golden L.E.A.F. (the N.C. State Tobacco Trust Fund), and Cherokee Preservation Foundation, a nonprofit organization that supports Cherokee cultural heritage. Funds were awarded through the Western Regional Educational Service Alliance (WRESA) and Southwestern Planning Commission, a regional planning and economic development organization.

Since the initial funding award, BalsamWest has assisted the schools by raising additional regional philanthropic funding of ***$3.0 million*** to connect the schools and community college sites in the remote mountain communities of Robbinsville and Cashiers, NC. The cost to build the long distances through steep, rocky terrain to connect schools in these communities was high. The ***Cherokee Preservation Foundation*** and the ***Carlton Family of Cashiers, North Carolina*** stepped forward to fund these long builds and donate connecting fiber-optic strands to the schools in Robbinsville and the schools and library in Cashiers.

The schools' ownership of their own fiber network ensures their complete control and freedom to choose the "content" they wish to receive, and from whom they receive it. Content includes Internet Service, High Definition Video programming of all types, digital surveillance, video and voice telephony, dedicated data transmission, distance learning programming, and shared computing applications. BalsamWest does not require the schools to purchase these content services through BalsamWest.

Funding has been awarded and construction is underway to connect 51 of the 70 K-20 schools. These schools will be connected by during the next 12 months. There are 19 institutions left to connect, and at this time, and BalsamWest and the ASAP partners and community leaders are working hard together to raise funding to connect the remaining sites and link them all together on their own fiber-optic ring

The patient capital Drake Enterprises and the Eastern Band of Cherokee Indian invested into the region has already begun to pay tremendous rewards to all. The business model of BalsamWest FiberNET is sustainable. Its model of providing metropolitan area wholesale pricing for dark fiber, colocation, and high-capacity services within and through the region to carriers, local enterprises, Internet Service Providers, and other providers and consumers of electronic content is working. BalsamWest was cash flow positive in its first six months of operation, and again at the end of 2006. This is quite an accomplishment for any new venture, much less a venture involved in rural fiber-optic deployment and wholesale transport services in a mountainous rural area. There was a large amount of pent-up demand for connectivity within the region, as well as to connect to a major metropolitan area and Tier One Internet PoP. As with any new venture, there will be ups and downs, but with the continued support of so many people who are so dependent upon the success of this venture for future prosperity and enhancements to quality of life, it will not fail.

Since September, 2003:

- ✓ BalsamWest has deployed over 300 miles of underground fiber optic cable in the southern Appalachian region of Western North Carolina, eastern Tennessee, and northern Georgia, meeting the critical need of the region for affordable access to reliable, high-performance state-of-the art fiber-optic broadband infrastructure, deploying underground for maximum network reliability, as well as to protect the beauty of the region.

- ✓ BalsamWest is providing direct access to lease strands of fiber in its fiber-optic network under operating or capital leases.

- ✓ BalsamWest is providing high-performance transport circuits at ***wholesale carrier pricing, comparable to metropolitan areas of the US*** where there is plentiful infrastructure and competitive wholesale pricing.

- ✓ BalsamWest offers a guarantee on service reliability comparable to the largest major US carriers serving worldwide businesses.

- ✓ BalsamWest is now serving local ISPs (Internet Service Providers), an electric cooperative, schools, hospitals, libraries, a mental health system, county and municipal governments, a regional economic development commission, and, soon, two real estate resort developments for Triple Play and Quadruple Play services. BalsamWest has interconnected with a major U.S. carrier and is working with an incumbent local exchange carrier to provide dark fiber and high-capacity circuits to expand service within the area.

- ✓ BalsamWest has adopted an "empty pipe" policy. BalsamWest offers only fiber and "empty pipes" (dedicated transport circuits used to transport "content" – voice, data, video and computing applications) within and through the area. BalsamWest does not sell content. Local content providers serve rural communities with low populations.

- ✓ Through the outreach of the Southwestern Community College, and representatives of Drake Software, the Eastern Band of Cherokee Indians, BalsamWest FiberNET, and all of the collaborators in the venture, rural EDCs (Economic Development Commissions") are becoming aware of their new capability to offer facilities in the region, connected to Atlanta, as locations for electronic operations of Atlanta companies. Rural EDCs are reviewing or deploying technology incubators connected directly to the fiber backbone for use by small businesses and entrepreneurs in conducting electronic businesses.

BalsamWest's future plans include:

- Expansion of the middle mile network into more mountainous rural communities in rural western North Carolina, eastern Tennessee, northern Georgia, and western South Carolina.
- Connectivity to first tier research universities in surrounding areas for access to additional distance learning programming for rural schools, and to allow technology businesses to tap the intellectual and research capabilities of these institutions.
- Connectivity of rural municipal and county government sites in the BalsamWest footprint to provide a secure and reliable, low-cost, high-performance, private government network so that they may obtain the same cost savings and benefits as the schools served by WNC EdNET.
- Connectivity from the Eastern Band of Cherokee Indians Tribal Goverment offices in Cherokee, NC to the remote reservation lands of Snowbird in the isolated, remote area of Graham County, NC.
- Additional connectivity for Triple Play and Quadruple Play content providers into the region for new residential developments connected on fiber.
- Raise awareness of public and private sector organizations concerning the broadband infrastructure situation of rural areas of the U.S. with significant terrain barriers
- Seek ongoing collaboration and support for network expansion to nearby metropolitan areas through the Southern Appalachian Mountain region.

- From a policy and support standpoint, even with our own regional fiber network we, or the companies to which we provide service, must interconnect to the existing telephone networks in each community for local loop access to premises. This means that we must lease the last mile of copper from the large incumbent carriers. Congress must ensure that our company, or the carriers and ISP companies to which we provide service, have access to the existing incumbent local network at reasonable rates.

- Mr. Chairman, we cannot rely on the old monopoly companies to drive broadband to rural America, especially in areas like the southern Appalachians that are so underserved. We have heard promise after promise from these companies "Just change this law or regulation and we will deploy advanced networks." If we want broadband available quickly to rural America, we must open the networks to competitors so that we can insure everyone in our area is able to benefit from the good work that we have begun. BalsamWest is an OPEN network offering services to all. We welcome competitive access to our network to increase the number and kinds of advanced services that can be provided at competitive pricing to residents, businesses, government, education, and health care institutions. To build on the good work begun by BalsamWest and ensure that the maximum benefit is obtained by our region, we, and other carriers, require competitive access to local incumbent networks in the communities we serve.

Summary

BalsamWest's work has indeed been a stunning success for everyone who has been touched by the high-performance network, thanks to the support of so many mountain people and organizations throughout the Southern Appalachian Region. Much work remains to be done, and more challenges will surely arise. But, the resiliency and commitment of the mountain people to independently solve challenges will overcome future challenges. The work of this isolated remote mountain region is a perfect example of the entrepreneurial spirit that has spurred so much innovation in the U.S. By collaboration and pooling of capital, resources, and expertise, the isolated mountain communities of this region are networked together, and can network small business and entrepreneurs together on an ultra-high-speed superhighway of virtually unlimited capacity. The region can distribute jobs, access to information, expertise, and resources across an entire regional network, and the network has been connected to one major metropolitan area trading center, with more to come. Through bold action and perseverance in the face of seemingly impossible challenges, the Southern Appalachian Mountain region is stepping forward as a new entrant into the global new economy, bringing a wealth of innovative ideas, products and services to the world at large – which is now at their fingertips.

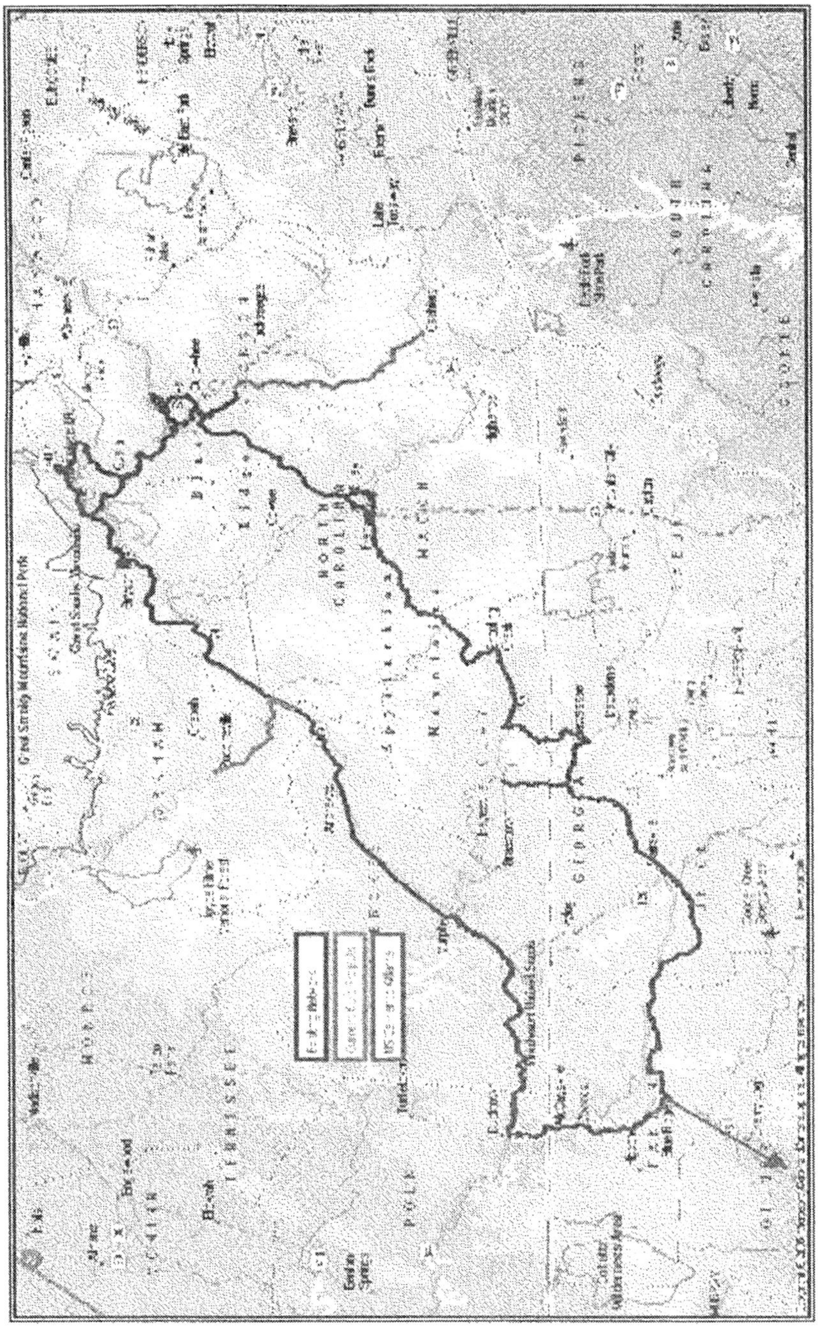

SOUTHERN APPALACHIAN COMMUNITIES CURRENTLY SERVED BY BALSAMWEST FIBERNET

Western North Carolina:

- **Jackson County:** Cashiers, Cullowhee, Dillsboro, East Laport, Gay, Glenville, Greens Creek, Sylva, Tuckasegee, Webster and Wilmot
- **Qualla Boundary of Eastern Band of Cherokee Indians**: Cherokee
- **Swain County:** Almond, Bryson City, Nantahala , Talc Mountain, and Whittier
- **Graham County:** Robbinsville and Tulula, (crosses Fontana Lake on the railroad's bridge trestle)
- **Cherokee County:** Andrews (adjacent to Andrews airport), Hothouse, Marble, Murphy, Ranger, Suit, Tomotla and Topton
- **Macon County:** Cullasaja, Franklin, Iotla, Pumpkintown, Rainbow Springs, Union, and traverses Chunky Gal
- **Clay County:** Hayesville, Shooting Creek, and adjacent to Ridges Country Club

Eastern Tennessee:

- Polk County: Copperhill, Ducktown and Isabella

Northern Georgia:

- Fannin County: Blue Ridge, McCaysville and Morganton
- Towns County: Friendship, Jacksonville, Hiawassee, Young Harris
- Union County: Blairsville

TESTIMONY OF T. RUSSELL SHIELDS
CHAIR, YGOMI LLC

BEFORE THE SUBCOMMITTEE ON RURAL
AND URBAN ENTREPRENEURSHIP
OF THE HOUSE COMMITTEE ON SMALL BUSINESS
U.S. HOUSE OF REPRESENTATIVES
ON
MAXIMIZING THE VALUE OF
BROADBAND SERVICE TO RURAL COMMUNITIES

May 9, 2007

Testimony of T. Russell Shields
Chair, Ygomi LLC
Before the Subcommittee on Rural and Urban Entrepreneurship
of the House Committee on Small Business
U.S. House of Representatives
on Maximizing the Value of
Broadband Service to Rural Communities

May 9, 2007

Chairman Shuler, Ranking Member Fortenberry and Members of the Subcommittee, my name is Russ Shields. I am chair of Ygomi LLC. It is a privilege to have the opportunity to speak to you today.

Ygomi is a U.S.-based holding and operating company with a 37-year track record of building companies that successfully deliver innovative software and services to meet essential business needs. We are known for the imaginative application of information technologies that improve the lives of people throughout the world.

We currently operate four information and communications technology companies that are developing and commercializing leading-edge solutions in such areas as wireless digital signal processing software, vehicle telematics, and technical support for multi-location enterprises using distributed call centers. Ygomi is headquartered in Oak Brook, Illinois. Ygomi and its companies – SEI, Verety, Connexis, and ArrayComm – serve leading corporations around the world, with more than 1,200 employees across the U.S., Europe, and Asia.

Today's hearing on "Maximizing the Value of Broadband Service to Rural Communities" is important to the Ygomi family of companies because broadband technology touches all of them. For instance, the expansion of high-speed internet connectivity to rural areas in the U.S. has provided unique opportunities for companies like ours to invent, improve, and evolve new and economic business solutions for our customers. In addition, we have been able to provide broadband services directly to our employees improving their quality of life and the economic viability of their communities.

Service firms, including call center providers in the U.S., are increasingly facing issues of labor shortages. The deployment of broadband and high-speed interconnections has provided a new source of reliable, talented work-at-home labor to U.S. service businesses, helping us to keep those jobs here in America and to provide higher quality service to our customers.

North Dakota is one state that has benefited from superior broadband deployments. Verety, one of the Ygomi companies, provides remote order taking services for a number of McDonald's U.S. restaurants, helping them improve speed of service, order accuracy, and customer satisfaction. Verety is able to deliver consistently high-quality service because the availability of broadband services gives us access to an exceptional talent pool in North Dakota that is able to work from home. We take great pride in bringing job opportunities to people in rural states like North Dakota who cannot or prefer not to work outside their homes. Verety's work-at-home opportunities offer greater workforce participation, a high degree of flexibility, and an enhanced quality of life. This approach attracts a higher caliber work force with more education and work experience than traditional centers can. We can offer attractive career opportunities, which in turn means a more stable workforce.

Today, we have approximately 150 employees in the areas surrounding Fessenden, Rugby, Steele, and Wishek, North Dakota. We have a dynamic workforce that includes farmers, stay-at-home mothers, retirees, people with disabilities, and people who care for elderly or disabled family members. We expect the number of work-at-home employees to increase exponentially in the years to come. We provide each work-at-home employee with a computer and software; internet DSL connection, telecommunications equipment, paid training, and web and phone-based support. The at-home broadband service is provided by Qwest, United Telephone, and a variety of rural telephone companies and cooperatives. In addition, our employees and their families can use the computer and internet connections for themselves when not working. Our employees like their no-commute savings and the convenience of flexible work shifts of 2 to 10 hours. It is interesting to note that more than two-thirds of Verety's work-at-home employees had no access to broadband for their families before coming to work for us.

Currently, many of our operations and training managers work from their homes. We have also identified several key positions that can be migrated to the work-at-home model and which represent career paths for some of our current employees. These include Senior Technical Analysts, Quality Assurance Analysts, Knowledge Managers to support customer service agents, Inside Sales representatives who nurture client accounts, and Process Engineers who help us improve the way we do business. These positions require highly skilled people with college degrees and relevant work experience, and they can earn significantly more than minimum wage.

Other Ygomi companies are also taking advantage of broadband deployments in rural areas. Broadband is at the core of many of our business models – past and present. For instance, ArrayComm, widely acknowledged as the world leader in commercializing multi-antenna signal processing software for wireless communications systems, developed software that is currently deployed worldwide including in Australia, Canada, Africa, China, Japan, Europe, and the Middle East. Its technology greatly enhances the capacity and improves the economics of wireless mobile broadband services. This technology will help extend the reach of broadband services in both rural and urban areas around the world. Connexis is working in partnership with vehicle manufacturers worldwide to provide safety-related data connectivity for all vehicles, with particular emphasis on areas with little or no cellular coverage. We foresee broadband and other advanced communications technologies making a major contribution to improve road safety for people in rural areas.

We have all read Thomas Friedman's book, *The World is Flat*. In fact, it mentions our effort for McDonald's in North Dakota. I believe that the vision of a "flat world" is becoming truer each day. The days of Private Branch eXchanges (PBXs) and Automatic Call Distributions (ACDs) are no more. Broadband services like DSL and broadband applications like VoIP are bridging the gap and helping give smaller, more isolated communities more access to the world and vice versa. Like Verety, companies in the Telework Coalition are helping to realize the benefits of broadband deployment in some rural areas. The same can be done elsewhere in the U.S., if the proper incentives are provided. But it will take more than just broadband to bring real benefits and economic opportunities to these areas. It will require a new way of thinking about the workplace and technological innovation.

Broadband is just one way to deliver these benefits that works better in some areas than others. We encourage and support the deployment of other communications technologies that enable creative solutions and public and private partnerships in rural areas to ensure that the next generation of benefits are available to everyone no matter where they live. We are working with organizations like TIA to promote access to affordable and advanced communications services, to limit regulations, minimize disruption to competitive market forces, and promote the use of broadband in government services, public safety, education, teleworking, and healthcare.

Increased global competition requires a more flexible labor environment. As a privately-held company, Ygomi is able to take a patient, long-term approach to profitability. We have the flexibility to think about the future and the technologies and applications that will be needed 10 or more years from now. Despite this flexibility, we still face challenges. For instance, new technologies increasingly change the way we conduct business and the ways our employees choose to work. Employee attitudes about work change, and to succeed, we have to respond creatively to new and evolving employee needs. Non-traditional work arrangements and alternative compensation structures are needed. This century's workplace will not be in just one geographic location, but will be anywhere and any time. Among other things, the workplace will be shaped by the increasing participation of older workers who remain active longer, working parents who want to balance their home and work lives, including the impact of commuting on the environment, and the desire to work where we live and to be able to choose places to live without being restricted to places with large employment centers.

Further, companies like ours who are working on emerging technologies and building businesses in rural areas will still need to seek out favorable business incentives that take into account investments in the community, equipment, wages, and training. This will require the implementation of national policies that encourage investment in new and diverse communications technologies, the promotion of competition as a means of facilitating universal deployment of broadband technologies, and fiscal incentives for broadband deployment. There are also elements in current tax and labor laws that, while appropriate for urban offices, make it difficult for companies to meet the needs and interests of employees working from their farm homes.

In conclusion, I would like to commend you and your staff for the holding this important hearing and for your efforts to help maximize the value of broadband to rural America. Thank you for the opportunity to testify before you today. I am pleased to answer your questions.

National Farmers Union

Testimony of Russell Kremer

Before the
U.S. House of Representatives
Committee on Small Business
Subcommittee on Rural and Urban Entrepreneurship

Concerning Broadband Technology & Economic Development in Rural America

Wednesday May 9, 2007
Washington, D.C.

STATEMENT OF RUSSELL KREMER

PRESIDENT, MISSOURI FARMERS UNION

BEFORE THE U.S. HOUSE OF REPRESENTATIVES SUBCOMMITTEE ON RURAL AND URBAN ENTREPENEURSHIP

CONCERNING BROADBAND TECHNOLOGY & ECONOMIC DEVELOPMENT IN RURAL AMERICA

MAY 9, 2007

Chairman Shuler, Congressman Fortenberry and members of the subcommittee, thank you for the opportunity to testify today. My name is Russ Kremer, and I am the president of the Missouri Farmers Union. Today I am here on behalf of the National Farmers Union (NFU), our nationwide organization representing family farmers, ranchers, fishermen and rural residents. In addition to running a diversified family farm in Osage County, Missouri, I am a board member of Cooperation Works, a national network of cooperative developers. I appreciate the opportunity to highlight the importance of accessible and reliable broadband service to the producers and communities of rural America.

The future of rural America, particularly family farmers and ranchers, depends on high-speed access to the internet. In 2005, the National Agricultural Statistics Service (NASS) conducted a study on farm computer usage and ownership. While the results showed that 51 percent of U.S. farms had internet access, further investigation uncovered that dialup was the most common method of accessing the internet with 69 percent of U.S. farms. It is encouraging that more farmers and ranchers gain computer accessibility each year, either through ownership/leasing of computers or through community programs. However, it is alarming that the vast majority of them must do so at the slowest connection speed possible.

NFU supports efforts to provide competitively priced, high-speed broadband internet access for rural America. We urge collaborative efforts and public-private initiatives that leverage internet based technologies and use the internet to improve communications, reduce costs, increase access and grow farm businesses for producers and their cooperatives. Illinois based NOW Wireless, LLC, and the Missouri Farmers Union, in response to the demand for affordable, modern telecommunications access for farmers and residents living in remote areas, helped establish USA Broadband, LLC (USAB). USAB has partnered with subscriber based cooperatives and developed successful networks that are making this access possible. To date, USAB is the premier provider of high-speed wireless broadband internet, voice communication and video services to rural communities. The company maintains its focus on providing a superior broadband product backed with exceptional customer care.

In June 2006, the Eastern Illini Electric Cooperative, IlliCom Telecommunications and USAB joined forces to offer CONXXUS, a broadband internet company that focused on providing a wide range of high-speed broadband service to rural residents within Eastern Illini Electric Cooperative's 10-county service area (6,000 square miles and 240,000 homes and businesses). USA Broadband President, Barry Goodwin, was quoted as saying, "CONXXUS has a solid foundation already established in the Paxton, Illinois community. We will continue to provide these same quality services and outstanding customer relations, while effectively expanding our service territory to include all of east-central Illinois. CONXXUS will provide that technology and offer unprecedented access to such things as Virtual Private Networks (VPN), distance learning opportunities, telecommuting, telemedicine, and complete connectivity throughout whole towns." Eastern Illini Electric Cooperative is a member-owned rural electric cooperative based in Paxton, Illinois that supplies energy solutions to approximately 11,000 members. IlliCom Telecommunications, a subsidiary of Eastern Illini, is a provider of high-speed wireless internet services, website design and hosting, video/internet services for multiple dwelling units, and dial-up internet services in eastern Illinois.

The internet is a necessary tool for farmers and ranchers, who will be at an economic and competitive disadvantage if unable to use the same high-speed internet connections that are available to other small businesses around the country. Farmers and ranchers rely on the internet to check weather, market and crop reports and search for suppliers of feed and equipment. Furthermore, family owned farming businesses need real-time access to online banking, accounting, order fulfillment and freight forwarding. Given the current economic climate, it is imperative that producers devote as much time as possible to marketing their products and exploring new markets. The ability to conduct financial transactions online would save individual producers hours of administrative work and translate into tremendous financial incentives at the farm level.

Access to broadband is imperative if we are to renew the rural landscape. Retaining and attracting aspiring young farmers and rural entrepreneurs is essential to securing a sustainable rural economy. NFU has developed local, community based cooperatives that produce and process food with the wholesomeness and integrity that consumers demand. Broadband services provide the link between these collective entrepreneurial businesses in rural areas to consumers throughout the country.

NFU's *www.e-cooperatives.com* is the world's first innovative portal to directly locate and buy quality food products, plus other goods and services, online from hundreds of U.S. agricultural producers and their co-ops in rural America. Both producers and consumers are able to access the *www.e-cooperatives.com* database via searchable categories for American family farms, ranches, cooperatives and rural businesses according to location, type of business, growing practices, available products, specialty goods and niche items. The E-Commerce Timeline Learning Model guides producers through every step of the process, from planning to web development and product design to marketing. This site originated as a technical assistance project of NFU and initially funded in part by grants from USDA's Rural Development agency. By eliminating the digital divide and providing more rural areas with high-speed internet access, we can help producers market and sell their quality products and educate consumers about the value of family farms and ranches.

Providers do not deny that there is a need for expansion into less urban areas, but investing in rural America needs to be attractive to investors and providers. Technical assistance is important. Broadband access cannot be expanded without providing the support infrastructure necessary to make it successful. High-speed access is a wonderful tool, but significant administrative challenges of getting into the homes of agricultural producers must be addressed. A single broadband company with a monopoly would have no incentive to maintain the same level of service offered to subscribers in heavily populated areas or to provide any service to sparsely populated areas; mergers and consolidations that remove or limit competition in rural markets should not be permitted. It is important to get an accurate portrayal of where the neediest areas are and how to provide broadband technology in those areas. Investigation into broadband access in rural areas reveals that most beneficiaries reside in larger towns.

Service providers tend to exercise great selectivity in the sectors of rural America in which they invest. As producers, we face great obstacles in attracting adequate financing and equity investment for broadband buildout in under-served areas. One solution is to establish federal incentives or tax credits to investors who supply equity to rural broadband initiatives in under-serviced areas. Similarly, federal loan guarantees or a reduction of the required equity match of the USDA Rural Utilities Service (RUS) program from 20 percent to one percent could make expansion projects more attractive and viable. I encourage you to evaluate these respective funding commitments and give consideration to decreasing the amount of money from state resources and mandate the federal government provide the maximum amount of funding possible.

The House Agriculture Subcommittee on Specialty Crops, Rural Development and Foreign Agriculture recently held its own hearing on rural broadband programs operated by USDA's RUS. The outcome of the hearing was a resounding message that access to broadband is the limiting factor in the economic growth of our rural communities. I encourage this subcommittee to work with your counterparts in the agriculture sector to find the most efficient and affordable way to provide rural America with reliable broadband access.

We believe that USDA's RUS is the appropriate agency to help expand technology and keep it up to date. RUS is an excellent example of how the federal government, rural cooperatives, nonprofit associations, public bodies and for-profit entities can work together to shrink the urban-rural divide. Funding of RUS programs should be increased above current levels and account for the additional staff that will be needed to accommodate the needs of rural citizens.

Similar to the first days of electricity, rural America is being left behind. It should be a national priority to include rural areas of the country in broadband buildout. Rural access to advanced telecommunications provides tools for enhanced medicine and education. The slow pace of rural broadband expands the educational divide in our country.

Reliable access opens the door for distance learning opportunities in rural schools. As a family farmer, I am greatly concerned that lack of high speed access is driving the younger generation out of rural America and into more urban environments. Better broadband means a better place to live, increased entrepreneurship, and retention of young people. High-speed internet connections make it more appealing for other businesses to enter rural areas that would otherwise refrain from establishing commercial outlets.

We applaud the efforts of the farmers and ranchers who have taken the initiative to work towards a reliable and affordable expansion of broadband technology. In fact, the genesis for the most successful cooperative efforts can be found on the farms and ranches of rural America, not in the bustling urban areas of the nation. I thank you for including an agricultural representative on today's panel and welcome any questions you may have.

TESTIMONY OF

BRENT CHRISTENSEN

CHRISTENSEN COMMUNICATIONS
MADELIA, MINNESOTA

UNITED STATES HOUSE OF REPRESENATIVES

SMALL BUSINESS SUBCOMMITTEE on

RURAL and URBAN ENTREPRENEURSHIP

MAY 9, 2007

Mr. Chairman and members of the committee, thank you for the opportunity to testify this morning. My name is Brent Christensen. I am the Vice President and General Manager of Christensen Communications Company. We are an independent, local exchange telecommunications carrier located in Madelia, Minnesota. It would be easier to tell you that we are a telephone company, but quite frankly, that is no longer an accurate description. I also have the privilege of serving as the chairman of the Legislative Policy Committee for the Organization for Promotion and Advancement of Small Telecommunications Companies (OPASTCO).

Our company was founded in 1903 by 48 local people who wanted state of the art telecommunications. One of those original 48 was my great-great-grandfather, Henry Joerg (the local blacksmith and saddle maker). This original group approached the owner of the local flour mill, C. S. Christensen (my other great-great-grandfather), and asked him to purchase 25% of the original stock. Over the years, my family acquired more and more stock, and today my father is the sole stockholder.

In 2006, we stopped using the Madelia Telephone Company name altogether. Christensen Communications Company better reflects what our business has become. We had customers who never thought of us when they needed their computers repaired or even for high-speed Internet.

We are very integrated in our community. We employ six people, not counting my parents and me. All but one of our employees reside in the community. We encourage our staff to be active in the community. Our employees are or have been volunteer Firefighters, EMTs, and Boy and Girl Scout leaders. We are active in the Chamber of

Commerce and other civic organizations. I currently serve on the Madelia Public School Board and am Vice President of the Chamber of Commerce. I also previously served as Mayor of Madelia and president of the Madelia Development Corporation.

I am here today to talk about broadband's impact on rural communities and Madelia in particular. We started providing broadband in 2000. We didn't start by putting a business plan together and figuring out how much money we could make. We started offering DSL because it is important to the economic survival of our community. We entered into the DSL business because Marv Davis needed it.

Marv, and his son Will, own Davis Sales and Service, a local Polaris dealer. We had been offering dial-up Internet service for a few years, as was a competitor. They told me that Polaris had changed the way they sold their snowmobiles, watercraft, and ATVs. Warranties were now issued over the Internet. When a customer came in to buy a snowmobile, the Davis' would fill out the customer information online and print off a warranty application. Once the customer had signed the document, the Davis' would scan the document and transmit it back to Polaris over the Internet. The problem was that dial-up was too slow for this process and their dial-up connection would frequently time out and they would have to start over. This was a frustrating process for the Davis' and their customers. In the end, if we didn't solve the problem, the Davis' would sell fewer Polaris', and it would severely impact their business.

I did some research on different solutions that would work with our network. We bought some equipment and got DSL service to the Davis'. The entire process took about 20 days. We didn't do a business case first, we didn't have to go through corporate bureaucracy, we just got a new service to a customer who needed it. The hard part was figuring out what to charge. It took us awhile, but in the end we settled on a rate that was both fair to the consumer and eventually recovered our costs.

When I was in high school, I worked at the telephone company as the summer help. My grandfather was president of the company at the time. I remember the two of us walking back to the office one day and him telling me how important the telephone company was to the community and how we had a responsibility to provide the best service possible. Back then it meant providing quality, reliable telephone service. Today it means much more. Today we have to provide state of the art communications for the survival of our small town. Madelia is like a lot of other towns our size, and in many ways like the communications industry itself. We are in competition with other communities in our area. We are in competition for industry and people. As a community we have to leverage our assets to develop our economy. Communications is one of those assets. Because of our communications infrastructure, we can market our town to telecommuters, small businesses, and others who do not depend on a specific location to conduct their business.

A good example of this is the House of Print. They are a local printing company that was started in the 1960s by a company that owned two daily newspapers in towns about twenty miles from Madelia to the north and south. Both papers needed to replace their printing facilities, and instead of each buying new presses, they built a new printing operation in Madelia which is halfway between the two. Today The House of Print prints newspapers for 100 daily and weekly newspapers.

The House of Print was our third DSL customer. Before they received high speed Internet, drivers would have to bring floppy disks containing the newspaper pages to Madelia to be printed. Proofs would have to be either faxed or mailed to customers, and the company was very geographically limited. Our high speed Internet allowed them to expand their customer base and increase their business. They have literally brought in millions of dollars of new business because of their high-speed Internet connection.

The House of Print is no longer geographically limited. Today they can bid for printing jobs online, allow the customer to upload data, proof the job on line, and mail the finished product directly from their facility. They have the advantage of being centrally located in the United States, which makes shipping their finished product that much easier. The House of Print has expanded significantly as a direct result of the Internet. They have added or upgraded their printing presses and expanded their building facilities.

The House of Print has become very dependent on the Internet for their business. So much so that they have had to add a redundant Internet connection. While we are now the only dial up Internet provider in Madelia, we have two high speed Internet competitors, Midwest Wireless, a cellular provider, and Comcast Cable. The House of Print gets their redundant Internet connection from Comcast.

As a small, rural company, we face many challenges providing state of the art communications. We have to provide all of the same services as the larger companies. This gives us a good understanding of our customers.

A good example of this is Farmers State Bank in Madelia. They are a locally owned independent bank. They compete against the Madelia branch office of a much larger bank. Our high-speed Internet connection has allowed Farmers State Bank to offer a full line of Internet banking services. I personally balance my checking account online and have even started paying my bills online. These services have kept Farmers State Bank competitive with other banks in our area.

Companies like Christensen Communications look to Congress for leadership on issues and programs that give us the opportunity to thrive, and in turn, keep our customers and community thriving. We ask Congress to continue to support a strong and viable Universal Service Fund (USF). The USF is the most important federal program for our continued success. Congress and the Federal Communications Commission needs to support the reform of the intercarrier compensation regime by implementing the Missoula Plan, which was developed by a broad cross section of the telecommunications industry. And Congress needs to support programs at the Agriculture Department's Rural Utilities Service and the Small Business Administration that help small businesses like mine.

We face many challenges in this industry, which directly affect our company and our ability to provide the advanced services our customers need to stay competitive in their businesses, like Davis Sales and Service, the House of Print, and Farmers State Bank.

I would like to thank you for the opportunity to testify today, and I would be happy to answer any questions you may have.

CONNECTED NATION

Testimony of Brian R. Mefford, CEO of Connected Nation
Before the U.S. House of Representatives
Committee on Small Business

Subcommittee on Urban and Rural Entrepreneurship

"Maximizing the Value of Broadband Services to Rural Communities"

Wednesday, May 9, 2007 - 10:00 a.m.
Room 2360 of the Rayburn House Office Building

Testimony of Brian R. Mefford, CEO of Connected Nation
Before the United States House of Representatives
Committee on Small Business
Subcommittee on Rural and Urban Entrepreneurship
Rep. Heath Shuler, Chairman
"Maximizing the Value of Broadband Services to Rural Communities"
Wednesday, May 9, 2007 - 10:00 a.m.
Room 2360 of the Rayburn House Office Building

Chairman Shuler and Members of the Committee:

Thank you for the opportunity to speak with you today regarding the important issue of expanding broadband and related technology to the rural areas of the United States.

Connected Nation is a national non-profit organization dedicated to closing the digital divide. Connected Nation is the parent company of ConnectKentucky, our Kentucky-based organization that has served as the "demonstration project" for Connected Nation. It is the "Kentucky story" that I'm here to share with you today.

Four years ago, Kentucky faced the same challenges that are all too common in rural states and communities across the country. The Commonwealth was struggling to use technology-centered solutions to address traditional challenges related to education, healthcare, and government services.

On the economic development front jobs in manufacturing, farming, and mining were leaving the state at an alarming pace, with little evidence that lost opportunities were being replaced with new technology-centric ones.

The results of Kentucky's technology troubles were not hard to identify. Kentucky consistently ranked low among states in terms of broadband availability and usage, as well as the number of high-tech companies doing business in the Commonwealth. Further, college graduates were leaving in droves, creating a troubling "brain drain" effect.

As we surveyed the landscape for answers, the reality of the situation was troubling indeed. We realized that the foundation of broadband infrastructure was not adequate for creating solutions that could address the challenges of a new day; not adequate to provide widespread access to telemedicine, distance learning and e-government; not adequate for growing or attracting entrepreneurs and industry; not adequate for providing more opportunities to our farm communities and their families where children were leaving their rural roots, never to return.

It was at this economic crossroads that Kentucky determined to become aggressive in addressing technology shortcomings to ensure that communities could come to thrive in this new environment – and ConnectKentucky, Kentucky's tech-based economic development partnership was born. In an unprecedented alignment of public and private interests, ConnectKentucky, an independent non-profit organization has seized upon the promises of the knowledge-based economy.

May 9, 2007 Small Business Committee Testimony 2

Immediately, ConnectKentucky set out to identify the root cause that had resulted in a lackluster technology picture for the state. It was clear that the inadequacy of Kentucky's broadband infrastructure could be traced to much of the state's inability to compete in areas important in the knowledge-based economy. Broadband infrastructure had been built into the state's more populous areas, leaving more rural areas unserved. The lack of service not only created the well-termed "digital divide" for rural residents, it also made it impossible to develop statewide policies that depended upon access to broadband.

Further, it was discovered that broadband availability was only half the problem. The remainder of the challenge related to the actual use of broadband-related technology. Any resulting turn-around strategy had to be comprehensive in nature: addressing both sides related to broadband *availability* and the *use* of broadband and related technology.

Next, the organization identified the barriers that were inhibiting broadband availability and use. In terms of availability there were a series of issues that needed to be addressed. First, very little data existed to allow us to identify the true extent of the broadband gaps in Kentucky. Providers didn't know, policy makers didn't know and communities themselves didn't know. Second, the regulatory environment was creating uncertainty among the provider community and causing a pull-back effect that ensured that investments weren't being made in more risky areas. Third, the cost of entry into rural communities was prohibitive for telecommunications providers of all types.

Challenges related to the use of technology included: lack of appreciation for the value of technology at the household level, lack of cohesive interest in technology at the local level, and lack of state initiatives to encourage awareness and build interest in technology at the state level.

Under the structure of a public-private partnership, ConnectKentucky developed a plan to address Kentucky's broadband challenge. The plan provided the direction for:

1) Gaining a better understanding of where we were – in the form of broadband inventory maps. The maps would help promote current service while also identifying the gaps that existed. Data layers would provide additional household information while identifying existing public assets, such as water towers, that could be used to extend broadband coverage;
2) Creating market intelligence at a local level to help providers identify investment opportunities and to effectively lower the cost of market entry. Household and business surveys would assist providers in better targeting rural investments;
3) Establishing grassroots technology leadership teams at the county level to create local technology strategies across multiple sectors including: local government, business and industry, education, healthcare, agriculture, tourism, libraries, and community-based organizations. The local teams would generate and aggregate demand by identifying ways to better leverage technology in local communities; and
4) Developing public-private initiatives that could promote the value of technology, improve technology literacy, and drive adoption among households.

Working in conjunction with Kentucky Governor Ernie Fletcher, the plan was fully developed, funded and launched as Kentucky's *Prescription for Innovation* in October 2004. The plan called for full broadband availability by the end of 2007; dramatically improved use of computers and the Internet; the creation of a meaningful online presence for all local communities; and the establishment of eCommunity Leadership Teams in all 120 counties.

As a result of the implementation of the *Prescription for Innovation*, Kentucky has experienced a technology turnaround. Due to the public-private approach and the very tactical nature of the plan, we have been able to bring all parties to the table to ensure that Kentucky could leapfrog its previous poor standings.

Through the work of ConnectKentucky and its partners, Kentucky's *Prescription for Innovation* has led to the following successes during the last two years:

- Kentucky is recognized as the **national leader** in technology acceleration with the *Prescription for Innovation* repeatedly acknowledged as the national model for states;
- Broadband inventory maps have been created for the entire state, promoting current coverage and allowing providers to better target unserved areas;
- **Broadband availability has increased from 60 percent to 92 percent** of households able to subscribe (on track to reach 100% by the end of 2007), representing 504,000 previously unserved households and more than 1.2 million residents that can now access broadband;
- **Broadband use at home has increased 73 percent**, a rate that has led the nation;
- Broadband use among Internet connected businesses rose from 65 percent to 85 percent;
- **Home computer ownership grew by 20 percent** while the national average rose by 4 percent;
- More than **$650 million in private capital has been invested in Kentucky** (unprecedented);
- Nearly **2,000 home computers have been distributed** to the homes of underprivileged Kentucky students through the No Child Left Offline program;
- **eCommunity Leadership Teams have been established in every Kentucky county** creating grassroots technology growth plans across nine sectors;
- More than **70 percent of Kentucky counties now operate or are in the process of constructing a meaningful web presence** for e-government and online citizen services, up from about 30% just two years ago;
- **22,000,000+ positive media impressions** have covered Kentucky technology growth; and

At an increasing rate, companies are locating to Kentucky, entrepreneurs are developing businesses in Kentucky, and jobs are growing in Kentucky because the Commonwealth now has the technology infrastructure and an increasing technology-savvy workforce to support business growth.

Over the last two years, more than 14,500 total technology jobs have been created in Kentucky[1]. Perhaps the most appropriate place to isolate and measure the direct employment impact of broadband expansion efforts is in the Information Technology (IT) sector. During the same two year period, **in the IT sector alone, Kentucky jobs have grown at a rate 31 times the national growth rate: 3.1 percent for Kentucky versus 0.1 percent nationally.**

[1] Bureau of Labor Statistics (BLS) for two year period beginning January 2005 through December 2006. Includes jobs created in the following NAICS sectors: information; finance; professional, science, and technical; management; and healthcare. Sectors are comprised primarily of high tech jobs and all jobs within these sectors are "technology based". Other sectors include additional technology jobs; however, these jobs are aggregated with other non technology jobs, such as in the manufacturing sector. As BLS does not disaggregate these jobs, they could not be included in the figure above, which results in an understatement in the reporting of technology jobs.

The closing of the digital divide is already yielding dividends in the quality of life for Kentuckians. Computer literacy has increased, the number of high tech jobs has increased, and Kentucky communities are enjoying the return of their children. Consider these developments in higher education related to how Kentucky has addressed the "brain drain" challenge that all states face:

- Today, 86 percent of all Kentucky graduates remain in Kentucky to live and work—a 18 percent increase since 2000;
- Since 2000, there has been a 50 percent increase in the number of out-of-state students who remain in Kentucky;
- For those graduates who came in as Kentucky residents, 95 percent of them now stay; and
- The percent of doctoral degree students who stay in Kentucky has nearly doubled (27 percent to 52 percent).

Today in Kentucky entrepreneurs are thriving; businesses of all sizes are finding an environment ripe for growth; rural communities are finding ways to diversify and provide attractive opportunities for their children; primary schools and universities are connected as never before, providing content and curriculum never before possible. Kentucky has developed a statewide eHealth plan that recently received federal funding as part of an aggressive Medicaid Transformation program. In short, as the broadband challenge has been addressed a strong foundation was established to allow for technology-centric solutions to flourish.

Kentucky as a microcosm has demonstrated the importance of the national broadband discussion and the relevance of technology to America's ability to compete. Based on our experience in Kentucky, we know that technology diminishes the significance of distance. In the past, opportunities to thrive have depended largely upon one's proximity to major markets. Technology has made the distance factor irrelevant. In other words, with the availability of cutting edge technology, entrepreneurs can thrive just as well in rural America as places such as New York or Los Angeles. Technology has become the great equalizer for individuals and communities alike – creating opportunities, fueling better education, higher quality healthcare, and better quality of life – regardless of where an individual or community happens to be located.

This same dynamic however represents both a huge opportunity and major threat for the United States. Other countries have invested in broadband towards achieving universal access – and like Kentucky, they have managed to leapfrog their previous standings to become a competitive force. It is the hope of Connected Nation that this Congress can call the country to arms on this issue by conveying the true sense of urgency for action. The nation needs a comprehensive approach that is good for our markets, our entrepreneurs and our communities. No doubt, it is a challenge of historic proportion. Just as previous times called for a national response to the needs for railroads, highways, electricity, and telephone service – the broadband challenge calls for an aggressive and comprehensive response to ensure that America remains the dominant leader in the global economy.

Thank you Mr. Chairman.

Background

CONNECTED NATION.

Connected Nation, Inc. is a national non-profit organization known for its ability to close the digital divide. Through its partnerships, programs, and policies Connected Nation makes technology work for previously underserved communities and markets, improving community life and economic development while enhancing markets for technology providers. Connected Nation's proven methodologies are delivering dramatic results that translate into more efficient public services and enhanced quality of life. Connected Nation's work in Kentucky, ConnectKentucky, has been identified as a national model for the expansion of broadband.

Connected Nation's proven methodologies enable comprehensive technology expansion efforts that effectively enhance the supply of available broadband while dramatically increasing demand through state and local grass roots awareness/adoption campaigns. Connected Nation specializes in increasing technology access and literacy towards greater digital inclusion for all. This technology expansion improves economic development, healthcare, education, and public safety; and provides a better way of life for Americans.

Charting the course for the United States' technology-centric future, Connected Nation creates partnerships between the public and private sectors. These partnerships encourage cooperation for mutually beneficial purposes – making the cost of technology expansion go down and the demand for technology go up. Our comprehensive approach to technology expansion works for communities and markets.

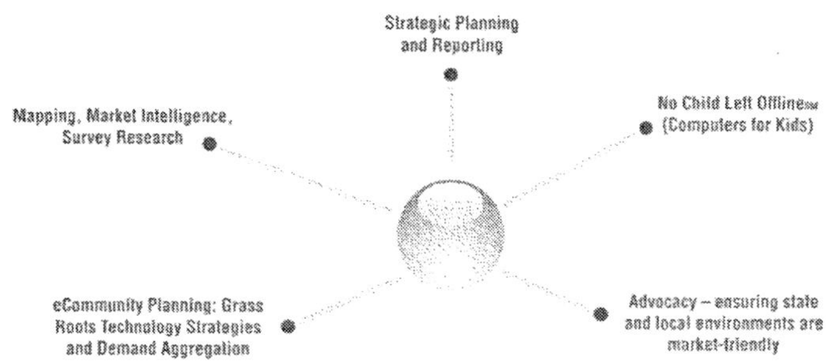

May 9, 2007 Small Business Committee Testimony 6

Research and Mapping

Connected Nation's broadband inventory maps are industry leaders. These GIS maps create an inventory of existing broadband services based on provider deployment data. This analysis effectively helps broadband providers to more effectively target their build out resources. Connected Nation's market intelligence (maps, survey data and grassroots demand aggregation) benefits companies by causing the cost of doing business to go down and the ease of doing business to go up.

This broadband inventory map is publicly available and based on provider deployment data.

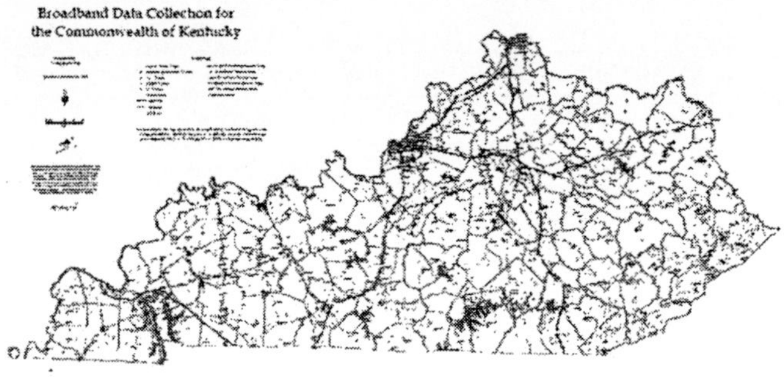

Beyond accurately measuring the inventory of broadband services, Connected Nation's research measures other important items related to the expansion of broadband. For example, What are the consumer barriers to broadband? Or, How do businesses use broadband?

Based on these findings, programs can be developed that encourage digital inclusion. For example, our research indicated that while industry assumed that the monthly fee was a primary barrier to the adoption of household broadband the lack of a computer at home ranked even higher. We developed No Child Left Offline as a partnership based solution. No Child Left Offline has facilitated cooperation among private partners, corporate foundations and state governments to place computers and printers into the homes and schools of disadvantaged children.

May 9, 2007 Small Business Committee Testimony 7

Connected Nation's Impact

Connected Nation's model is based on a simple premise that technology can be good for communities and markets. Comprehensively engaging both supply and demand realities is the best plan for success. The results from ConnectKentucky confirm the strength of Connected Nation's model.

Launched in 2004, Kentucky's *Prescription for Innovation* is a comprehensive plan to accelerate technology statewide, particularly in the areas of broadband availability and computer literacy and use. ConnectKentucky is implementing this initiative which maintains four key objectives for impacting statewide technology-based economic development:

- Full broadband deployment;
- Dramatically improved use of computers and the Internet by all Kentuckians;
- A meaningful online presence for all Kentucky communities, to improve citizen services and promote economic development through e-government, virtual education, and online healthcare; and
- Local technology leadership teams in every community to develop and implement technology growth strategies for local government, business and industry, education, healthcare, agriculture, libraries, tourism, and community-based organizations.

As identified by the *Prescription for Innovation*, technology can dramatically expand economic development opportunities and improve the quality of life for Kentuckians. With expanded technology, opportunities are within reach, such as:

- Developing a competitive economic advantage for attracting today's high-tech jobs to replace the decline of traditional manufacturing jobs;
- Residing in one of Kentucky's rural communities and succeeding in a career that formerly required moving to a major metropolitan area;
- Better and less expensive healthcare; and
- An education that prepares Kentucky's children to prosper in a globally networked world.

To fully address each of these opportunities and to ensure that Kentucky provides an increasingly attractive environment for technology expansion, ConnectKentucky employs a comprehensive approach that has been identified as a national leader and a model program for the rest of the country to follow.[2] Last year, ConnectKentucky received the U.S. Economic Development Administration's 2006 Excellence in Innovation Award.

[2] ConnectKentucky has been cited as a national best practice by: the US Economic Development Administration, the U.S. Government Accountability Office of Congress, the White House Office of Technology, Federal Communications Commission, Appalachian Regional Commission, USDA Rural Utilities Service, Congressional Research Service, Center for Digital Government, Southern Growth Policies Board, Communications Workers of America, Rural Telecommunications Congress and numerous states across the nation.

What Is ConnectKentucky?

ConnectKentucky connects people to technology in world-altering ways: improving the lives of the formerly disconnected; renewing hope for previously withering rural communities; driving increases in the number of tech-intensive companies and jobs; and nurturing an environment for lifetime learning, improved healthcare, and superior quality of life. Through its partnerships, programs and policies ConnectKentucky makes technology work for previously underserved communities and markets, improving community life and economic development while enhancing markets for technology providers.

ConnectKentucky works with supply and demand realities in a manner that respects communities and gets results. ConnectKentucky is engaged with all 120 Kentucky counties, local business and community leaders, and private sector technology companies to facilitate comprehensive technology expansion efforts that both enhance the supply of broadband-related technology and create demand by catalyzing and delivering grassroots awareness, literacy and use of technology.

Impact of ConnectKentucky

Through the work of ConnectKentucky and its partners, Kentucky's *Prescription for Innovation* has led to the following successes during the last two years:

- Kentucky is recognized as the **national leader** in technology acceleration with the *Prescription for Innovation* repeatedly acknowledged as the national model for states;
- **Broadband availability has increased from 60 percent to 92 percent** of households able to subscribe, representing 504,000 previously unserved households and more than 1.2 million residents that can now access broadband;
- **Broadband use at home has increased 73 percent**, a rate that has led the nation;
- Broadband use among Internet connected businesses rose from 65 percent to 85 percent;
- **Home computer ownership grew by 20 percent** while the national average rose by 4 percent;
- More than **$650 million in private capital has been invested in Kentucky** (unprecedented);
- Nearly **2,000 home computers have been distributed** to the homes of underprivileged Kentucky students through the No Child Left Offline program;
- **eCommunity Leadership Teams have been established in every Kentucky county** creating grassroots technology growth plans across nine sectors;
- More than 70 percent of Kentucky counties now operate or are in the process of constructing a meaningful web presence for e-government and online citizen services. Two years ago, only one-third of Kentucky counties had a website, and many of these were not functional;
- 22,000,000+ positive media impressions have covered Kentucky technology growth; and
- **Kentucky is on track to be the first state with 100 percent broadband coverage.**

ConnectKentucky's Economic Impact: A Case Study

At an increasing rate, companies are locating to Kentucky, entrepreneurs are developing businesses in Kentucky, and jobs are growing in Kentucky because the Commonwealth now has the technology infrastructure and an increasing technology-savvy workforce to support business growth. **On track to become the first state with 100 percent broadband coverage with nation-leading increases in broadband use at home and work.**

Over the last two years, more than 14,500 total technology jobs have been created in Kentucky[3]. The most appropriate place to isolate and measure the direct employment impact of broadband expansion efforts is in the Information Technology (IT) sector. During the same two year period, **in the IT sector alone, Kentucky jobs have grown at a rate 31 times the national growth rate: 3.1 percent for Kentucky versus 0.1 percent nationally.**

Chart 1

Kentucky Growth vs. National Growth in Information Technology Jobs

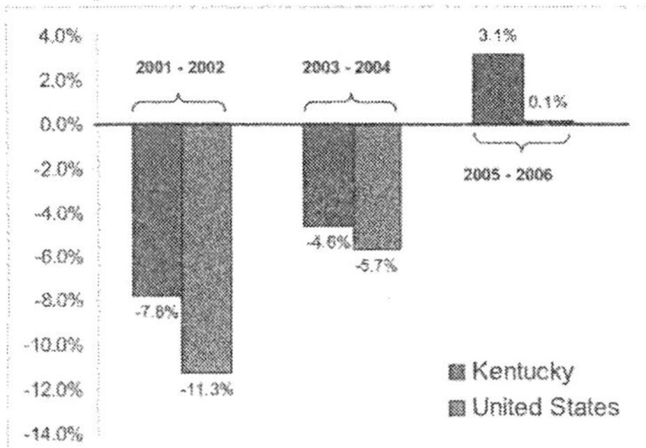

During the first two years of the *Prescription for Innovation*, Kentucky IT jobs grew by 3.1 percent, outpacing national growth by 31 times.

Connected Nation's Kentucky engagement, ConnectKentucky, has been recognized as a national leader by: the U.S. Government Accountability Office of Congress, the White House Office of Technology, US Economic Development Administration, Federal Communications Commission, Appalachian Regional Commission, USDA Rural Utilities Service, Congressional Research Service, Center for Digital Government, Southern Growth Policies Board, Communications Workers of America, Rural Telecommunications Congress and numerous states across the nation.

[3] Bureau of Labor Statistics (BLS) for the two year period beginning January 2005 through December 2006. Includes jobs created in the following NAICS sectors: information; finance, professional, science, and technical; management; and healthcare. These sectors are comprised primarily of high tech jobs and all jobs within these sectors are "technology based". Other sectors include additional technology jobs; however, these jobs are aggregated with other non technology jobs, such as in the manufacturing sector. As BLS does not disaggregate these jobs, they could not be included in the figure above, which results in an understatement in the reporting of technology jobs.

Reversing the "Brain Drain" in Kentucky

The closing of the digital divide is already yielding dividends in the quality of life for Kentuckians. By closing the digital divide, computer literacy has increased, the number of high tech jobs has increased, and Kentucky communities are enjoying the return of their offspring. Consider these developments in higher education related to how ConnectKentucky has helped Kentucky address the "brain drain" challenge that all states face:

- Today, 86 percent of all Kentucky graduates remain in Kentucky to live and work—a 17 percent increase since 2000.
- Since 2000, there has been a 50 percent increase in the number of out-of-state students who remain in Kentucky.
- For those graduates who came in as Kentucky residents, 95 percent of them now stay.
- The percent of doctoral degree students who stay in Kentucky has nearly doubled (27 percent to 52 percent).

Connected Nation's Legislative Agenda

Connected Nation provides the leadership that delivers technology for strong communities and open markets. Our work is predicated on the notion that there's no reason for anyone in America to be on the wrong side of the digital divide.
Here's why:

- Connected Nation's work has proven to be effective in state based engagements like ConnectKentucky.
- Each state has underserved communities that desperately need access to affordable and dependable broadband.
- National public and private entities are looking for a means of cooperating for our greater national good.

Therefore, Connected Nation encourages legislation that bridges the digital divide for all of America.

Connected Nation advocates a national legislative agenda that accomplishes the following:

- Provides solution for ubiquitous broadband deployment and increased adoption by encouraging and funding public-private partnerships at a state level;
- Establishes a grant program to enable each state to develop a comprehensive approach to broadband deployment while simultaneously driving broadband adoption and technology development at a local community level;
- Allows nonprofit organizations that have established a partnership with state government to apply for funding to:
    - Identify and map the gaps in broadband service – those areas without broadband availability – and then work collaboratively with all providers to fill those gaps in a manner that supports their business plans and works for communities;
    - Measure and track broadband and information technology use among citizens and businesses, investigate barriers to adoption at a local level, and provide market analysis for unserved areas;
    - Develop local technology planning teams with members representing a cross section of the community, including business, telecommunication labor, K-12 education, health care, libraries, higher education, community-based organizations, local government, tourism, parks and recreation, and agriculture;
    - Equip and facilitate local technology planning teams with the tools and resources to improve technology use within each sector; and
    - Establish effective programs to improve computer use and Internet access for disenfranchised populations.

This public-private partnership approach establishes the collaborative environment that encourages investment, drives technology adoption, and empowers grassroots-led community development and ultimately, strengthens America.

CONNECTED NATION℠

May 9, 2007 Small Business Committee Testimony 12

CONNECTED NATION℠

Technology for Strong Communities And Open Markets

Connected Nation offers valuable partnerships that enable technological and economic progress of historic proportion.

Please join Connected Nation as we close the digital divide in America.

444 North Capitol Street, Suite 224
Washington, DC 20001
1-877-846-7710
www.connectednation.com
info@connectednation.com

Statement of
William R. Deere
Vice President, Government Affairs
U.S. Telecom Association
to the
House Committee on Small Business
Subcommittee on Rural and Urban Entrepreneurship
May 9, 2007

Chairman Shuler, Ranking Member Fortenberry, members of the subcommittee: Thank you for this opportunity to appear before you today. I am Bill Deere, Vice President of Government Affairs for the USTelecom Association. I am pleased to appear before this subcommittee to discuss USTelecom's perspective on *"Maximizing the Value of Broadband Services to Rural Communities."* It is a timely moment for the subcommittee to hold this hearing.

USTelecom and its member companies are proud of the role we play connecting the country, and we wholeheartedly support the objective of ubiquitous, nationwide broadband. We were pleased to see "affordable broadband access for all Americans" as a component of Speaker Pelosi's Innovation Agenda. Similarly, the Senate Republican High-Tech Task Force is calling for policies that "promote widespread deployment and use of broadband technology." Broadband deployment and adoption *should* be *non*-partisan objectives, and we believe the Congress, the Federal Communications Commission and the Rural Utilities Service have vital roles to play in advancing these goals in rural America.

USTelecom represents innovative companies ranging from the smallest rural telecoms in the nation to some of the largest corporations in the U.S. economy. Our member companies offer a wide range of services across the communications landscape, including voice, video and data over local exchange, long distance, Internet and cable networks. USTelecom is the nation's most established – and largest – association representing rural telecom providers. The vast majority of our member companies are rural providers. They are small businesses serving small communities. They are proud members of these communities and deeply committed to their future development. What unites our diverse membership is our shared determination to deliver innovative voice, video and data services to the consumer—a commitment we know is shared by this subcommittee.

Regulatory Changes Have Spurred Broadband Deployment

The Federal Communication Commission's decisions that oriented the communications marketplace away from government-managed to market-based competition have resulted in an explosion of broadband coverage across the nation. In March 2002, the FCC clarified that high-speed cable-modem service is an information service not subject to unbundling and other Title II regulations of the Communications Act. In August 2003, the FCC exempted wireline fiber facilities from the Commission's unbundling requirements. In September 2005, the FCC clarified that wireline broadband Internet access service is also an information service not subject to unbundling and other Title II regulations of the Communications Act. These actions have accelerated broadband deployment in the United States from just over 4 million broadband lines in 2000 to just under 16 million broadband lines in 2002 to approximately 32 million lines in 2004 to almost 65 million lines in 2006. This demonstrates a direct correlation between the FCC's market-based policies and the explosion of broadband subscribers in the United States. The lack of regulation on wireless services also has permitted wireless broadband services to explode as well. In June of 2005, there were almost 380,000 wireless broadband subscribers; in June of 2006, there were more than 11 million. The Commission's recent video franchise order promises to further increase the demand for broadband service.

Internet access is available through DSL, or cable modem, or wireless, or satellite – and, increasingly, over power lines and municipal wi-fi systems. In fact, there are more than 1,270 broadband service providers in the U.S. today.

Against this competitive backdrop, North American telecommunications companies are projected to spend $70 billion on new infrastructure this year. The next wave of broadband innovation holds the promise of significant, life-enhancing advances from health care to the environment to education and to our economy. It is critical, as you know, that these opportunities be accessible in rural America, as well. Mr. Chairman, much has been made recently of new international broadband penetration rankings from the Organization of Economic Cooperation and Development (OECD). We have some issues of our own with our country's current ranking of 15th in the world. We feel it significantly undercounts, for example, connections in the U.S. business market. It certainly also under-values the markedly more intense facilities-based competition we have here in the U.S. But the most striking dissimilarity is that 10 of the 11 countries allegedly in front of us are significantly smaller than the U.S.—as diminutive as Norway, which is comparable in geographic size to New Mexico. A majority also have much smaller *populations*, including Iceland, an entire country that is comparable to the metro area of Naples, Florida. The exception is Canada, which is a country of vast geographic expanse. However, 80% of the population is clustered along the U.S. border. So the true broadband challenge before our country is precisely the challenge we are here today to discuss. How can we most efficiently work together to connect parts of the country where the marketplace alone is incapable of attracting the significant investment necessary to truly build a broadband nation?

USTelecom and our member companies are committed to furthering rural broadband deployment and believe that Congress can advance a number of initiatives that promote this goal.

Sustainable Universal Service

First, we must ensure a sustainable future for universal service, a program designed to increase access to telecommunications services nationwide and to maintain affordable rates in low-income and rural areas. USTelecom and our member companies have advocated that universal service should be reformed to create a strong and sustainable system that can provide affordable, reliable telecommunications for all Americans. The current funding system is eroding at a rapid pace requiring the current system to be reformed. The need for reform was underscored last week by the Federal-State Joint Board on Universal Services' recommendation to temporarily cap further growth of universal service funds going to wireless carriers and competitive service providers.

USTelecom supports this recommendation. Consumers would benefit from a temporary cap on universal service support because it reduces the burden they face from a rising contribution factor. A temporary cap on universal support for wireless carriers and competitive service providers would allow the Commission additional time to develop and implement much needed fundamental reforms. Doing nothing in the near term, which would only serve to raise the cost of communications services purchased by consumers, is simply not an option.

In the House, Representatives Rick Boucher (D-VA) and Lee Terry (R-NE) have recently re-introduced universal service reform legislation that they first proposed last year. This legislation is an important initiative to help preserve the future for universal telecommunications service and spur broadband deployment in rural areas and we appreciate the Congressmen's dedication to finding a sustainable, long-term solution. While there is broad recognition that action must be taken to reform universal service, it is vital that members of this Committee encourage the consideration of such reform legislation this year.

Tax Policies To Encourage Broadband Deployment

In addition, Congress can promote broadband deployment by permanently extending the Internet Tax Moratorium; allowing for faster depreciation of broadband equipment and fiber; and creating a tax credit for the deployment of broadband equipment and fiber.

Congress first passed the Internet Tax Freedom Act (ITFA) in 1998. The moratorium was extended by Congress in 2001 and 2004 and now expires Nov. 1, 2007. The moratorium needs to be permanently extended to ensure that this critical component of the American economy is not the target of excessive taxes imposed by state and local governments. If the moratorium is allowed to lapse, USTelecom members and their customers will face a significant tax increase for Internet access services.

USTelecom was joined by NCTA and CTIA in a letter to all House members in support of H.R.743, bipartisan legislation introduced by Representatives Anna Eshoo and Bob Goodlatte. I encourage all members of the Committee to consider cosponsoring this legislation and urge the House take up this important legislation before its expiration in November.

The RUS Broadband Program -- Modest Changes Could Produce Dramatic Results

In its relatively brief history, the Rural Utilities Service (RUS) broadband loan program has achieved some successes. But we believe with modest changes, largely based on the successful RUS telephone program, the program could accomplish even more.

Last week, USTelecom appeared before the House Agriculture Committee in order to make recommendations for inclusion in the Farm Bill that would advance our collective goal of helping the nation achieve universal broadband penetration:

1) Better target areas currently not served;
2) Enhance incentives for investment in the areas not served;
3) Expand program eligibility;
4) Improve processing at USDA; and
5) Explore public-private partnerships.

Revise the eligibility rules to better target areas not served

We believe the primary weakness of the current program is that it does too little for areas with no access to broadband. Although the nation is dotted with areas currently not served, the USDA Inspector General concluded the program's focus has shifted away from rural communities that would not, without government assistance, have access to broadband technology.

In revising eligibility rules, we believe the Agriculture Committee may need look no further than the RUS telephone program. This program has a 60-year record of success, and we believe it holds important lessons for broadband. In the telephone program, initial loans to areas with adequate, existing service are discouraged. In fact, the RUS administrator must issue a non-duplication finding prior to making such a loan. In the broadband program, a similar requirement would help direct funds to where they are most needed – those areas with no existing broadband service. Making loans for duplicative facilities and service, when other citizens in rural America reside in areas with no service at all, is not a good use of scarce government resources. In addition, the telephony program requires that service be extended to the widest practical number of users in the service area, avoiding a problem that has sometimes arisen in the broadband program, where service is only provided within town limits, but not to the surrounding county.

Enhance incentives for investment in areas not served

Providing broadband service in rural and remote areas is a challenging proposition. While the current practice of offering cost-of-money loans makes projects financially viable in some areas, other higher cost areas will require below-cost loans or a combination of loans and grants to make a costly infrastructure build feasible. This will become increasingly important as the program narrows to focus on areas with truly no access. Congress should encourage RUS to look at the unique needs of these areas and to enhance incentives for the private sector to act. Taxpayers will reap the benefits through loan repayments and tax revenues generated by broadband-driven economic development. We believe that taking these basic steps would increase the number of loan applications to areas with no service facing significant economic barriers to investment, such as low population densities or difficult terrain.

Expand eligibility to more applicants

We also believe steps should be taken to expand the number of companies eligible for broadband loans. When the broadband program was established, a provision was adopted prohibiting loans to telephone companies with more than 2% of the nation's access lines. This is counterproductive. Some USTelecom members serve rural areas that would otherwise qualify for broadband loans. For example, the FCC classifies Embarq as a rural carrier in 17 of the 18 states it serves, yet it is prohibited from applying for RUS broadband funds. Meanwhile, RUS is searching for more applications from carriers seeking to serve untouched areas. Again, if I might refer you to the successful, 60-year-old telephony program – it has never had a 2% restriction, and it has never suffered as a result. The emphasis in our view should be on the infrastructure needs of a community, not on the company willing to serve it.

Improve processing at USDA

USTelecom also advocates that steps be taken to improve processing of loan applications at USDA. At present, the broadband and telephony programs have access to a small number of attorneys in the Agriculture Department's general counsel office. This has created a bottleneck when legal decisions are needed and caused delays in processing loan applications—delays that too often put broadband deployment on hold in communities with no service.

Explore public-private partnerships

Finally, I direct the subcommittee's attention to the successful public-private partnership in Kentucky, driven by a non-profit organization called Connect Kentucky. Connect Kentucky has worked with the RUS broadband program, but has gone much farther than would have been possible with RUS alone. Its first objective was to map broadband availability in the whole state, something that no other state has done. Then it created technology teams in each community that lacked broadband. These teams looked at computer ownership, technological literacy, and other factors to increase demand for broadband. At the same time, the teams worked with broadband providers to match up new demand with new broadband deployments. By the end of 2007, Kentucky will go from having one of the lowest broadband subscription rates in the country to having broadband available to 100% of its households. That's impressive progress, and we think Congress might look to Connect Kentucky as a model for what works. In fact, we understand that Senator Durbin has recently introduced legislation that would create a national program based on the Connect Kentucky model.

Mr. Chairman, in closing, let me reiterate that it is critically important that rural areas be included in the nationwide drive for greater bandwidth capacity. This modernization of the nation's communications infrastructure will seed economic growth and expand opportunities ranging from telecommuting to distance learning to telemedicine. Mr. Chairman, nowhere in the nation do these advances hold more potential than in rural America.

We thank you for your invitation to appear today. USTelecom and its member companies look forward to working with the subcommittee and this Congress to achieve our shared objective of making broadband as ubiquitous today as electricity, water and telephone service. Broadband is an essential building block of every modern American community and we must make sure its many opportunities are accessible to all Americans. Thank you.

INDEX

A

B

C

D

E

F

G

H

P

Q

R

S